Protecting the Health and Well-Being of Communities in a Changing Climate

PROCEEDINGS OF A WORKSHOP

Leslie Pray, *Rapporteur*

Roundtable on Population Health Improvement

Roundtable on Environmental Health Sciences, Research, and Medicine

Board on Population Health and Public Health Practice

Health and Medicine Division

The National Academies of
SCIENCES · ENGINEERING · MEDICINE

THE NATIONAL ACADEMIES PRESS
Washington, DC
www.nap.edu

THE NATIONAL ACADEMIES PRESS 500 Fifth Street, NW Washington, DC 20001

This activity was supported by contracts between the National Academy of Sciences and AAMC, Aetna Foundation (#1002892), The California Endowment (#10003309), Dartmouth Hitchcock Medical Center, Exxon Mobil, General Electric Company (#10003080), Fannie Rippel Foundation, HealthPartners, Health Resources and Services Administration (DHHS-10003351), Kaiser Permanente (#10002957), Kresge Foundation (#10002910), Low Income Invest Fund, National Association of County & City Health Officials (#10003582), Nemours, New York State Health Foundation (#10002907), NIEHS (DHHS-10003294), NYU Langone School of Medicine, Office of the Assistant Secretary for Health (DHHS-10003388), Robert Wood Johnson Foundation (#10002897), Samueli Institute, and Wake Forest Baptist Medical Center. Any opinions, findings, conclusions, or recommendations expressed in this publication do not necessarily reflect the views of any organization or agency that provided support for the project.

International Standard Book Number-13: 978-0-309-46345-4
International Standard Book Number-10: 0-309-46345-9
Digital Object Identifier: https://doi.org/10.17226/24846

Additional copies of this publication are available for sale from the National Academies Press, 500 Fifth Street, NW, Keck 360, Washington, DC 20001; (800) 624-6242 or (202) 334-3313; http://www.nap.edu.

Printed in the United States of America

Suggested citation: National Academies of Sciences, Engineering, and Medicine. 2018. *Protecting the health and well-being of communities in a changing climate: Proceedings of a workshop*. Washington, DC: The National Academies Press. doi: https://doi.org/10.17226/24846.

The National Academies of
SCIENCES · ENGINEERING · MEDICINE

The **National Academy of Sciences** was established in 1863 by an Act of Congress, signed by President Lincoln, as a private, nongovernmental institution to advise the nation on issues related to science and technology. Members are elected by their peers for outstanding contributions to research. Dr. Marcia McNutt is president.

The **National Academy of Engineering** was established in 1964 under the charter of the National Academy of Sciences to bring the practices of engineering to advising the nation. Members are elected by their peers for extraordinary contributions to engineering. Dr. C. D. Mote, Jr., is president.

The **National Academy of Medicine** (formerly the Institute of Medicine) was established in 1970 under the charter of the National Academy of Sciences to advise the nation on medical and health issues. Members are elected by their peers for distinguished contributions to medicine and health. Dr. Victor J. Dzau is president.

The three Academies work together as the **National Academies of Sciences, Engineering, and Medicine** to provide independent, objective analysis and advice to the nation and conduct other activities to solve complex problems and inform public policy decisions. The National Academies also encourage education and research, recognize outstanding contributions to knowledge, and increase public understanding in matters of science, engineering, and medicine.

Learn more about the National Academies of Sciences, Engineering, and Medicine at **www.nationalacademies.org**.

The National Academies of
SCIENCES · ENGINEERING · MEDICINE

Consensus Study Reports published by the National Academies of Sciences, Engineering, and Medicine document the evidence-based consensus on the study's statement of task by an authoring committee of experts. Reports typically include findings, conclusions, and recommendations based on information gathered by the committee and the committee's deliberations. Each report has been subjected to a rigorous and independent peer-review process and it represents the position of the National Academies on the statement of task.

Proceedings published by the National Academies of Sciences, Engineering, and Medicine chronicle the presentations and discussions at a workshop, symposium, or other event convened by the National Academies. The statements and opinions contained in proceedings are those of the participants and are not endorsed by other participants, the planning committee, or the National Academies.

For information about other products and activities of the National Academies, please visit www.nationalacademies.org/about/whatwedo.

PLANNING COMMITTEE ON PROTECTING THE HEALTH AND WELL-BEING OF COMMUNITIES IN A CHANGING CLIMATE[1]

HENRY ANDERSON, Adjunct Professor, Occupational and Environmental Health, University of Wisconsin Department of Population Health Sciences; former Chief Medical Officer, Wisconsin Division of Public Health

PAUL A. BIEDRZYCKI, Director, Disease Control and Environmental Health, City of Milwaukee Health Department

JOHN BOLDUC, Environmental Planner, Cambridge Community Development Department

KATHY GERWIG, Vice President, Employee Safety, Health and Wellness, and Environmental Stewardship Officer, Kaiser Permanente

SURILI PATEL, Senior Program Manager, Environmental Health, American Public Health Association

LINDA RUDOLPH, Director, Climate Change and Public Health Project, Public Health Institute

Health and Medicine Division Staff

ALINA BACIU, Activity Co-Director
KATHLEEN STRATTON, Activity Co-Director
HOPE HARE, Administrative Assistant
ROSE MARIE MARTINEZ, Senior Board Director, Board on Population Health and Public Health Practice

Consultant

LESLIE PRAY, Rapporteur

---

[1] The National Academies of Sciences, Engineering, and Medicine's planning committees are solely responsible for organizing the workshop, identifying topics, and choosing speakers. The responsibility for the published Proceedings of a Workshop rests with the rapporteur and the institution.

ROUNDTABLE ON POPULATION HEALTH IMPROVEMENT[1]

GEORGE J. ISHAM (*Co-Chair*), Senior Advisor, HealthPartners; Senior Fellow, HealthPartners Institute for Education and Research

SANNE MAGNAN (*Co-Chair*), Adjunct Assistant Professor, University of Minnesota

PHILIP M. ALBERTI, Senior Director, Health Equity Research and Policy, Association of American Medical Colleges

TERRY ALLAN, Health Commissioner, Cuyahoga County Board of Health

JOHN AUERBACH, Executive Director, Trust for America's Health

CATHY BAASE, Chair, Board of Directors, Michigan Health Improvement Alliance; Consultant for Health Strategy, The Dow Chemical Company

RAPHAEL BOSTIC, Judith and John Bedrosian Chair in Governance and the Public Enterprise; Director, Bedrosian Center on Governance, University of Southern California

DEBBIE I. CHANG, Senior Vice President, Nemours

CHARLES J. FAZIO, Senior Vice President and Medical Director, HealthPartners

GEORGE R. FLORES, Senior Program Officer, The California Endowment

KATHY GERWIG, Vice President, Employee Safety, Health and Wellness and Environmental Stewardship Officer, Kaiser Permanente

ALAN GILBERT, Director of Global Government and NGO Strategies, GE Healthymagination

MARY LOU GOEKE, Executive Director, United Way of Santa Cruz County

MARTHE GOLD, Senior Scholar in Residence, The New York Academy of Medicine

MARC N. GOUREVITCH, Muriel G. and George W. Singer Professor of Population Health, Department of Population Health; Professor, Department of Medicine; Professor, Department of Psychiatry; Chair of the Department of Population Health, New York University School of Medicine

GARTH GRAHAM, President, Aetna Foundation

GARY R. GUNDERSON, Vice President, Faith Health, Wake Forest University

WAYNE JONAS, Executive Director, H & S Ventures

ROBERT M. KAPLAN, Professor, Stanford University

---

[1] The National Academies of Sciences, Engineering, and Medicine's forums and roundtables do not issue, review, or approve individual documents. The responsibility for the published Proceedings of a Workshop rests with the workshop rapporteur and the institution.

DAVID A. KINDIG, Professor Emeritus of Population Health Sciences; Emeritus Vice Chancellor for Health Sciences, University of Wisconsin–Madison

PAULA M. LANTZ, Associate Dean for Academic Affairs and Professor of Public Policy, University of Michigan

MICHELLE LARKIN, Associate Vice President, Associate Chief of Staff, Robert Wood Johnson Foundation

THOMAS A. LaVEIST, Professor and Chair, Department of Health Policy, Milken Institute School of Public Health, The George Washington University

JEFFREY LEVI, Professor, Milken Institute School of Public Health, The George Washington University

SARAH R. LINDE, Chief Public Health Officer, Health Resources and Services Administration

SHARRIE McINTOSH, Vice President for Programs, New York State Health Foundation

PHYLLIS D. MEADOWS, Senior Fellow, Health Program, The Kresge Foundation

BOBBY MILSTEIN, Director, ReThink Health

JOSÉ T. MONTERO, Director, Office for State, Tribal, Local and Territorial Support; Deputy Director, Centers for Disease Control and Prevention

MARY PITTMAN, President and CEO, Public Health Institute

PAMELA RUSSO, Senior Program Officer, Robert Wood Johnson Foundation

JAMES N. WEINSTEIN, CEO and President, Dartmouth-Hitchcock, Peggy Y. Thomson Professor, Evaluative Clinical Sciences, Dartmouth-Hitchcock Medical Center

Roundtable on Population Health Improvement Staff

ALINA BACIU, Director

CARLA ALVARADO, Program Officer (*from October 2017*)

KIMANI HAMILTON-WRAY, Senior Program Assistant (*from April 2017*)

DARLA THOMPSON, Program Officer (*until August 2017*)

ROSE MARIE MARTINEZ, Senior Board Director, Board on Population Health and Public Health Practice

ROUNDTABLE ON ENVIRONMENTAL HEALTH SCIENCES, RESEARCH, AND MEDICINE[1]

FRANK LOY (*Chair*), U.S. Representative to the 66th Session of the General Assembly of the United Nations

LYNN R. GOLDMAN (*Vice Chair*), Professor, Environmental and Occupational Health, Milken Institute School of Public Health, The George Washington University

HENRY A. ANDERSON, Adjunct Professor, Occupational and Environmental Health, University of Wisconsin Department of Population Health Sciences; former Chief Medical Officer, Wisconsin Division of Public Health

JOHN M. BALBUS, Senior Advisor for Public Health, National Institute of Environmental Health Sciences, National Institutes of Health

FAIYAZ BHOJANI, Chief Medical Officer, Global Manufacturing and Chemicals, Royal Dutch Shell

LINDA S. BIRNBAUM, National Institute of Environmental Health Sciences, National Institutes of Health

WAYNE E. CASCIO, Director, Environmental Public Health Division, National Health and Environmental Effects Research Laboratory, Environmental Protection Agency Initiative Office of Public Engagement

LUZ CLAUDIO, Associate Professor, Community Outreach & Education, Mount Sinai School of Medicine

DENNIS J. DEVLIN, Senior Environmental Health Advisor, ExxonMobil Corporation

RICHARD A. FENSKE, Professor and Associate Chair, School of Public Health and Community Medicine, University of Washington

DAVID D. FUKUZAWA, Managing Director, Health, The Kresge Foundation

BERNARD D. GOLDSTEIN, Consultant, Department of Environmental and Occupational Health, Graduate School of Public Health, University of Pittsburgh

RICHARD J. JACKSON, Professor and Chair, Department of Environmental Health Sciences, Fielding School of Public Health, University of California, Los Angeles

SUZETTE M. KIMBALL, Senior Advisor, Office of the Director, U.S. Geological Survey, U.S. Department of the Interior

---

[1] The National Academies of Sciences, Engineering, and Medicine's forums and roundtables do not issue, review, or approve individual documents. The responsibility for the published Proceedings of a Workshop rests with the workshop rapporteur and the institution.

JAY LEMERY, Assistant Professor of Emergency Medicine; President, Wilderness Medical Society, School of Medicine, University of Colorado Denver

MAUREEN Y. LICHTVELD, Professor and Chair, Freeport McMoRan Chair of Environmental Policy; Director, GROWH Research Consortium; and Director, Center for Gulf Coast Environmental Health Research, Leadership and Strategic Initiatives, Tulane University School of Public Health and Tropical Medicine; and Associate Director, Population Sciences, Louisiana Cancer Research Consortium

AL McGARTLAND, Office Director, National Center for Environmental Economics, Environmental Protection Agency

DAVID M. MICHAELS, Professor, Department of Environmental and Occupational Health, Milken Institute School of Public Health, The George Washington University

SUSAN L. SANTOS, Assistant Professor, School of Public Health, Rutgers University

KIRK P. SMITH, Professor of Global Environmental Health, School of Public Health, University of California, Berkeley

AGNES SOARES DA SILVA, Regional Advisor, Sustainable Development and Health Equity, Pan American Health Organization/World Health Organization

JOHN D. SPENGLER, Professor, Environmental Health and Human Habitation, Harvard T.H. Chan School of Public Health

G. DAVID TILMAN, Director, Cedar Creek Ecosystem Science Reserve; Regents Professor, Department of Ecology, Evolution and Behavior, University of Minnesota

JULI TRTANJ, One Health and Integrated Climate and Weather Extremes Research Lead, Climate Program Office, National Oceanic and Atmospheric Administration

PATRICIA VERDUIN, Chief Technology Officer, Global Research & Development, Colgate-Palmolive Company

DICK ZIMMER, former Congressman, Zimmer Strategies, Inc.

Roundtable on Environmental Health Sciences, Research, and Medicine Staff

KATHLEEN STRATTON, Director
HOPE HARE, Administrative Assistant
ROSE MARIE MARTINEZ, Senior Board Director, Board on Population Health and Public Health Practice

Reviewers

This Proceedings of a Workshop was reviewed in draft form by individuals chosen for their diverse perspectives and technical expertise. The purpose of this independent review is to provide candid and critical comments that will assist the National Academies of Sciences, Engineering, and Medicine in making each published proceedings as sound as possible and to ensure that it meets the institutional standards for quality, objectivity, evidence, and responsiveness to the charge. The review comments and draft manuscript remain confidential to protect the integrity of the process.

We thank the following individuals for their review of this proceedings:

PHILIP M. ALBERTI, Association of American Medical Colleges
WAYNE CASCIO, National Health and Environmental Effects
    Research Laboratory
ALLISON GOST, Maryland Institute for Applied Environmental
    Health
MAUREEN LICHTVELD, Tulane University

Although the reviewers listed above provided many constructive comments and suggestions, they were not asked to endorse the content of the proceedings nor did they see the final draft before its release. The review of this proceedings was overseen by **DEREK YACH** of The Vitality Group. He was responsible for making certain that an independent examination of this proceedings was carried out in accordance with standards of the National Academies and that all review comments were carefully considered. Responsibility for the final content rests entirely with the rapporteur and the National Academies.

Contents

Boxes, Figures, and Table

TABLE

1

Introduction[1]

On March 13, 2017, the Roundtable on Environmental Health Sciences, Research, and Medicine and the Roundtable on Population Health Improvement jointly convened a 1-day public workshop in Washington, DC, to explore potential strategies for public health, environmental health, health care, and related stakeholders to help communities and regions address and mitigate the health effects of climate change. Specifically, as Lynn Goldman, vice chair of the Roundtable on Environmental Health Sciences, Research, and Medicine, stated in her introductory remarks, the workshop objectives were (1) to receive an overview of the health implications of climate change; (2) to explore mitigation/prevention and adaptation/resilience-building strategies deployed by different sectors at various levels (e.g., local, national) and in various regions of the United States; and (3) to discuss aspects of collaboration on climate and population health issues among community-based organizations, health care systems, businesses, and public health and other local government agencies, along with lessons learned. These objectives, Goldman explained, were to be sought within the context of considerations of health equity, economic viability, social acceptability, political palatability, and regional fit. She

---

[1] The planning committee's role was limited to planning the workshop, and the Proceedings of a Workshop has been prepared by the workshop rapporteur as a factual summary of what occurred at the workshop. Statements, recommendations, and opinions expressed are those of individual presenters and participants, and are not necessarily endorsed or verified by the National Academies of Sciences, Engineering, and Medicine, and they should not be construed as reflecting any group consensus.

BOX 1-1
Statement of Task

An ad hoc committee will plan and convene a 1-day public workshop exploring the implications of climate change for population health and the potential strategies that public health, environmental health, health care, and related stakeholders can implement to help communities and regions address and mitigate health effects. The committee will develop the agenda and identify meeting objectives, select appropriate speakers, and moderate the discussions. The workshop will explore the perspectives of civic, government, business, and health-sector leaders, and highlight existing research, best practices, and examples that inform stakeholders and practitioners on approaches to support mitigation of and adaptation to climate change and its effects on population health. A Proceedings of a Workshop—In Brief and a complete proceedings based on the presentations and discussions at the workshop will be prepared by a designated rapporteur in accordance with institutional guidelines.

---

emphasized the "sharing" aim of the workshop and its "spirit of learning about best practices across the country."

Both the workshop objectives and agenda were developed by an ad hoc committee (the Statement of Task is presented in Box 1-1). The workshop agenda is provided in Appendix B. Biographies of the speakers and moderators are provided in Appendix C.

ABOUT THE ROUNDTABLES

As Goldman discussed in her welcome, since its first meeting in 1998, the Roundtable on Environmental Health Sciences, Research, and Medicine has addressed current and emerging issues in environmental health through discussions related to science, research gaps, and policy implications. The roundtable has held workshops on a range of issues of domestic and international importance, such as climate change, sustainable drinking water, ecosystem services, the health impact assessment of shale gas extraction, the science of obesogens, sustainable development, and data for environmental health decision making.

In his introductory remarks, George Isham, co-chair of the Roundtable on Population Health Improvement, said that since February 2013, the roundtable has been providing a trusted venue for leaders from the public and private sectors to meet and discuss the leverage points and opportunities arising from changes in the social and political environment. The roundtable's vision is of a strong, healthful, and productive society that cultivates human capital and equal opportunity. This vision, Isham continued, rests on the recognition that outcomes such as improved life expectancy, quality of life, and health for all are shaped by interdependent social, economic, environmental, genetic, behavioral, and health care factors and will require robust national and community-based policies and dependable resources. Isham remarked that this workshop reflected the roundtable's interest in the wide range of factors that impact human health.

ORGANIZATION OF THE WORKSHOP AND THIS PROCEEDINGS

The workshop was intended to provide an overview of the health implications of climate change, explore mitigation/prevention and adaptation/resilience-building strategies being implemented across the country, and discuss collaborative efforts on climate and population health issues. The workshop was not intended to describe or discuss in any detail the science of climate change, a topic of multiple consensus reports from the National Research Council, and from other expert sources (NRC, 2011, 2013; *The Lancet*, 2015).

Following introductory remarks by Goldman and Isham, Georges Benjamin of the American Public Health Association provided some further opening comments on the urgency of the issue. Keynote speaker Jonathan Patz then set the stage for the workshop with his presentation, "The Global Climate Crisis: Large Health Risks AND Opportunities."

The remainder of the workshop was organized into four panels, with each panel focused on a specific geographic region of the United States: Panel 1 provided regional perspectives from the South, Panel 2 from the Midwest, Panel 3 from the Northeast, and Panel 4 from the West. Each panel included two to four panelist presentations, followed by a discussion with the audience.

As Isham mentioned during his welcome, it has been a custom of the roundtable to conclude its workshops with reflections on the day's discussions. Thus, this workshop concluded with closing reflections by Sanne Magnan, Ray Baxter, and Frank Loy.

The organization of this summary parallels that of the workshop itself. Chapter 2, "Setting the Stage," summarizes Benjamin's opening remarks and Patz's presentation. Chapter 3, "Regional Perspectives from the South," summarizes the presentations and discussions that occurred during Panel 1. Chapter 4, "Regional Perspectives from the Midwest," summarizes the Panel 2 presentations and discussion. Chapter 5, "Regional Perspectives from the Northeast," presents the Panel 3 presentations and discussion; Chapter 6, "Regional Perspectives from the West," the Panel 4 presentations and discussion; and Chapter 7, "Reflections on the Day," summarizes closing remarks by Magnan, Baxter, and Loy and the open discussion that took place following their remarks.

2

Setting the Stage

Georges Benjamin's introductory remarks and Jonathan Patz's opening presentation set the stage for the remainder of the workshop.
This chapter summarizes Benjamin's remarks and Patz's presentations, with key points listed in Box 2-1.

Following his presentation, Patz participated in an open discussion
with the audience. A wide range of topics were addressed, including how
health professionals and urban planners can work together; implications
of geographic variation in indoor versus outdoor air pollution; tree canopy
spread as a mitigation strategy; the return on investing in green energy;
populations that are particularly vulnerable to the health effects of climate
change (children, elderly, the poor); the emerging issue of environmental
refugees; engaging leaders at the national and global levels; how to talk
about climate change; and political organizing and partnerships. This discussion is summarized at the end of this chapter.

THE FIERCE URGENCY OF NOW

We are now faced with the fact that tomorrow is today. We are confronted
with the fierce urgency of now. In this unfolding conundrum of life and
history there is such a thing as being too late. Procrastination is still the
thief of time. . . . We must move past indecision to action.

Martin Luther King, Jr.[1]

---

[1] Quote from the Rev. Dr. Martin Luther King, Jr.'s April 4, 1967, speech, *Beyond Vietnam:
A Time to Break Silence*, delivered at "a meeting of Clergy and Laity Concerned About

---

BOX 2-1
Key Points Made by Individual Speakers

- Climate change is not only a fiercely urgent challenge; it presents a tremendous opportunity for public health to become involved in solving a problem that affects every facet of human life. (Benjamin)
- Climate change encompasses more than rising temperatures. It also includes extreme hydrologic events, such as floods and droughts. A continuously growing body of evidence has associated these and other impacts of climate change with a range of health outcomes. (Patz)
- Climate change has been framed in multiple ways—it is not just an environmental issue, but also a national security, moral, and health challenge. (Benjamin, Patz)
- Huge opportunities for cost-effective, health-in-all-policies approaches to addressing climate change exist, particularly at the local level, in the energy sector, in food and agriculture, and in transportation and urban planning. (Patz)
- The most important messages to communicate when talking about climate change are: "Climate change is real. It's now. It's already happening. It's solvable." (Patz)

Georges Benjamin, executive director of the American Public Health Association (APHA), reminded the audience that just 2 weeks prior to this workshop, people were walking around Washington, DC, without coats, some in shorts, in contrast to the large snow storm forecast for the day after the workshop. "No one is going to tell you for sure that that's related to climate change," Benjamin said, but he encouraged workshop participants to keep in mind, over the course of the day, that such huge climate shifts and environmental changes are being observed right now, with many records being broken.

"Climate change is the most pressing public health problem that we have," Benjamin said. He cited several reasons why the health sector should be engaged. First is the global nature of the threat. "From a global perspective, it's huge," he said. Second, climate change affects every facet of our lives. Third, it represents an excellent opportunity to solve a problem using a public health approach, particularly a health-in-all policies approach. He emphasized the enormous opportunities for people from multiple sectors to work collaboratively on a problem that extends across a range of hu-

---

Vietnman at Riverside Church in New York City." See http://inside.sfuhs.org/dept/history/US_History_reader/Chapter14/MLKriverside.htm (accessed May 11, 2017).

man activities, from what we eat and how we grow our food to how we build our environment, how we live our daily lives, and what we burn for fuel. "Taking the politics out of this, simply to make sure that our planet and those of us that are on our planet actually survive, is an excellent opportunity," he stated.

APHA has been thinking about the impact of climate change on health since the early 1990s, according to Benjamin. He noted that several years ago, the APHA executive board declared "becoming the healthiest nation" a central challenge and that dealing with climate change aligns with this challenge.

Climate change also presents an enormous opportunity, Benjamin continued. "Certainly we know," he said, "climate change is here and impacting our health today." But some important policy makers do not support climate change, and some do not even believe it is real. This lack of support and disbelief create urgency, making climate change, Benjamin said, "an issue we must take on now." In his opinion, this urgency, in turn, creates an opportunity to "set the table from a health perspective" and for the health community to fill what he described as an "enormous void." Finally, Benjamin asserted, not only is action needed now, but timely action will also keep the issue in the forefront of the minds of the public, policy makers, and funders. He suggested four health-sector goals:

1. Shift the narrative around climate change to speak to people's values regarding their health and that of their families and loved ones. "People love polar bears," he said, "but they love their health and their family's health more."
2. Serve as a science and policy resource to inform sound policies and decision making and to evaluate the health equity impacts of those policies.
3. Influence and advance climate policies that would have the greatest impact on environmental justice and health equity outcomes.
4. Galvanize action to advance climate-healthy practices and behaviors that will make the greatest impact on public health and health equity.

The Year of Climate Change and Health

APHA declared 2017 the year of "Climate Change and Health," a declaration, Benjamin explained, with two goals. One is to raise awareness by educating people that climate change is a public health issue and not just an environmental issue. The second is to mobilize leaders who are interested in climate change, but have not started to take action yet.

The association is striving to meet these goals by proactively planning events, as well as adopting events, such as this Health and Medicine Division (HMD) workshop, that are aligned with the goals of 2017 as the Year of Climate Change and Health. Each month of 2017, in collaboration with its partners, APHA has been or will be focusing on a specific theme to raise visibility and build momentum. The year would conclude with the APHA Annual Meeting (theme: Climate Changes Health) in Atlanta.

In addition to this HMD workshop, another recent event adopted by APHA because of its alignment with APHA's work in climate change was the Climate & Health Meeting held in Atlanta on February 16, 2017. That meeting was held to fill the void of the deferred Centers for Disease Control and Prevention (CDC) Climate & Health Summit, Benjamin explained. He thanked former Vice President Al Gore for co-hosting the meeting, as part of Gore's Climate Reality Project, and acknowledged the other co-hosts as well: Harvard Global Health Institute, University of Wisconsin Global Health Institute, and University of Washington Center for Health & the Global Environment. In addition to these co-hosts, the meeting drew more than 50 partners. The meeting was funded by the Turner Foundation and held at the Carter Center. In fact, Benjamin noted, former President Jimmy Carter attended the meeting and greeted participants. The meeting drew about 340 local participants and thousands of remote participants. At one point, according to Benjamin, the meeting trended at #2 on Twitter, an indication, he said, that "a lot of people are interested in this."

Questions from the Audience:
Shifting the Narrative and Response Versus Mitigation

Following his remarks, Benjamin fielded two questions from workshop participants. First, George Isham asked Benjamin to expand a bit on the first of the four suggested health-sector goals that he listed, that is, to shift the narrative around climate change to speak to people's values around their health. Benjamin replied that the challenge is that people tend to think about what they are going to lose from climate change adoption, for example, by reducing use of fossil fuels. Instead, he said, "We ought to be talking more about what we gain, particularly from a population perspective." By acting now, he explained, people can make the environment cleaner and build jobs, ultimately at a reduced cost to society and for improved overall well-being. Even at an individual level, taking action now will make a difference, in his opinion. Based on his work as a former emergency physician, he foresees fewer people walking into emergency rooms with asthma attacks, fewer seniors being impacted by heat-related injuries, and less worry about flooding and vector-borne diseases associated with that flooding. He remarked that there have been yearly vector-borne infectious disease emergencies over the past several years and that the environment is changing in such a way that makes it more receptive to these diseases. "We can stop that," he said. "It may take years, but we can change it."

Benjamin was then asked by an unidentified member of the audience how the public health community is talking with decision makers about the need to respond to immediate, "every day" health impacts versus impacts that are anticipated in the future. Benjamin stressed the importance of taking care of the emergency "in front of us." He referred, again, to the snow storm expected to hit Washington, DC, on the day after this meeting, and stressed the importance of having a system in place that can respond to the coming storm. In his opinion, this means having a well-structured, well-funded emergency response system that can deal not only with situations such as snow removal, but also with seniors who will lose their heat and with other individuals who are at risk of being affected by severe weather.

That this same response capacity also helps with mitigation is important, he continued, because the core governmental public health system has been devastated over the past several years. For example, despite the possibility that Zika could return in the spring, at least in the south, mosquito control programs have languished. Now is the time to begin building those programs, he said. Benjamin cautioned that gutting the Environmental Protection Agency (EPA) regulations will put people at "extraordinary risk." These regulations were put in place because there were problems that needed to be fixed. "We're trying to get the average American to understand that," he said, so people can say to the resource allocators, "We need to have these systems, these programs, these regulations in place."

THE GLOBAL CLIMATE CRISIS: LARGE HEALTH RISKS AND OPPORTUNITIES[2]

Jonathan Patz's presentation set the stage for the remainder of the workshop, beginning with a description of the Intergovernmental Panel on Climate Change's future projections for average surface temperatures across the globe (IPCC, 2013). Patz compared 1986–2005, a period of time he described as a "low emissions scenario," which resulted in a 1 degree ($^{\circ}$C) warming by the end of the century (on average, worldwide), with 2081–2100, a period of time when, if "business as usual right now" proceeds, will see a warming of 7–8°C (again, on average across the world) (see Figure 2-1). He asked, "Now, what does this mean for health?" He answered, "This is why we're here—the so what?" He reiterated Benjamin's key point that climate change is not only a major health risk, but also actions against climate change represent a huge opportunity for health.

---

[2] This section summarizes information presented by Jonathan Patz, University of Wisconsin–Madison.

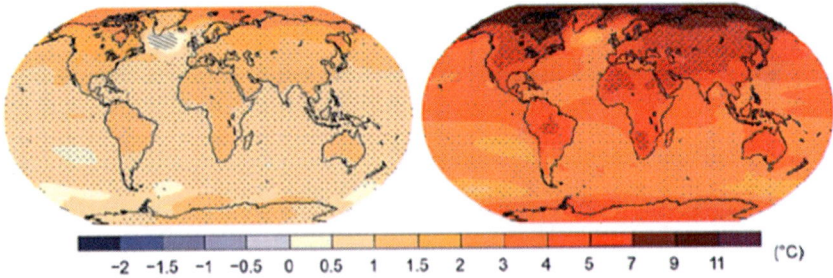

FIGURE 2-1 Change in average earth surface temperatures: The projected difference in increases in average earth surface temperatures between 1986–2005 and 2081–2100.
SOURCES: Patz presentation, March 13, 2017, adapted from *Climate Change 2013: The Physical Science Basis*; IPCC, 2013.

Health Effects of Climate Change

Patz emphasized that climate change involves not only temperature rise and sea level rise, but also hydrologic extremes (i.e., more droughts and more flooding, major blizzards and snowstorms). As Benjamin also noted, Patz referred to the record number of consecutive hot days in Washington, DC, in February 2017, followed by a blizzard expected the day following this workshop. These changes, he said, "cut across all sorts of health outcomes that we know are climate sensitive."

Such outcomes include heat stress and cardiovascular failure associated with heat waves; air pollution and aeroallergens (i.e., chronic obstructive pulmonary disease, asthma, and other respiratory diseases); infectious diseases, particularly vector-borne diseases (e.g., malaria, dengue, encephalitis, hantavirus, Rift Valley fever); water-borne diseases (e.g., cholera, *cyclospora*, *cryptosporidiosis*, *campylobacter*, *leptospirosis*); water resources and food supply problems (e.g., malnutrition, diarrhea, toxic red tides); and mental health outcomes and environmental refugees (i.e., forced migration, overcrowding, infectious diseases, human conflicts) (Patz et al., 2000). Patz cautioned that it is very difficult to attribute environmental refugees to climatic events, but suggested that a drought or sea-level rise that forces population movement could be the "iceberg under the tip of the iceberg as far as the extent of impact."

In addition to these listed highly climate-sensitive health outcomes, he noted that new findings emerge "every day." For example, recent studies out of Southeast Asia show increases in hypertension and preeclampsia associated with sea-level rise and salinization of freshwater aquifers.

Patz went on to discuss some of these health outcomes in more detail, beginning with outcomes associated with extreme heat events in New York City (NYC).

Temperature Rise: Examples of Health Impact

Currently, NYC experiences, on average, 13 summer days that are 90°F or hotter. It is predicted that this number will triple, from 13 to 39 days, by 2046–65 (Patz et al., 2014). This predicted tripling, Patz noted, holds for all of the eastern cities he and his colleagues have examined thus far. "This is a major concern," he said. "We know that people die in heat waves."

Not only do people die in heat waves, but crops become damaged. It has been predicted that many places worldwide will experience unprecedented summer heat in the future, with many regions experiencing a 90 percent or greater probability of having the hottest summer on record by 2080–2100 (Battisi and Naylor, 2009). Patz remarked that, based on this prediction, today's 840 million people at risk of hunger could double by mid-century.

While crops are damaged by rising temperatures, some plants may actually thrive in warmer temperatures and with higher CO_2 levels. This is "bad news," Patz said. He explained how work conducted by Lewis Ziska, U.S. Department of Agriculture, has shown that ragweed pollen season has been increasing over the past few decades and, therefore, children with asthma are at risk for a longer period of time (Ziska et al., 2016).

Regarding Zika virus, which Benjamin had noted, Patz mentioned that a new study shows that the risk for transmission of Zika was higher in 2015–2016 than it has been in more than 60 years, based on climate suitability of the *Aedes aegypti* mosquito (Caminade et al., 2016). This is not surprising, in Patz's opinion, based on what is known about dengue fever—the two viruses are very similar, with the same mosquito vector. In Southeast Asia, where dengue fever occurs every year and with an expected seasonal peak, the couple of years that have experienced major epidemics have coincided with very strong El Niño weather events (i.e., 1997 and 2009) (van Panhuis et al., 2015). The 1997 El Niño was the strongest in recent history, Patz noted, until 2015. Based on what he and colleagues have observed with respect to surface temperatures in South America associated with the 2015 El Niño, Patz suspects that it had at least something to do with the recent emergence of Zika in South America.

Hydrological Extremes: Examples of Health Impact

Patz reiterated that climate change encompasses more than global warming—it also includes extremes in the hydrological cycle. According to the U.S. Global Change Research Program, Patz said, "In the future, when it rains, it will pour," because hot air holds more moisture. These rains affect public health through water contamination. A hard rain, combined with a system that is unable to handle the storm water, can lead to combined sewage overflow events. In an analysis of future precipitation intensity in Chicago, Patz and colleagues predicted a doubling of combined sewage overflow events by 2050 because of the increased rainfall and runoff (Patz et al., 2008).

In Syria, in contrast, decreasing rainfall, coupled with an increasing temperature, has led to extreme drought. Patz explained that before the civil war, which led to hundreds of thousands dead and millions of refugees, Syria experienced the most severe drought in the instrumental record (Kelly et al., 2015). While the impact of the drought on the refugee population is hard to pinpoint, to the extent that migration from rural to urban locations was three to four times its normal level, coupled with food prices being "through the roof," all in the midst of civil unrest, the drought could have had indirect, destabilizing, and enormous effects. According to Patz, potential effects like these are one reason why President Obama began talking about climate change as a national security issue.

Changing the Framing of Climate Change

In addition to President Obama's reframing of climate change as a national security issue, there have been efforts to reframe climate change in other ways as well, Patz continued. He referred to Benjamin's remarks on the present opportunity to reframe climate change as a health issue, and not just an environmental issue.

Climate change has also been framed as a moral issue (i.e., the "Pope Francis effect"). Ten years ago, Patz and colleagues (2007) published a cartogram (a map combining statistical and geographic data) showing which countries were producing the most CO_2, with the United States as the top emitter. Since then, Patz noted, China has surpassed the United States in CO_2 emissions. Yet, for the most climate-sensitive diseases (i.e., malaria, malnutrition, diarrheal disease), the greatest initial impact is in sub-Saharan Africa and India. The fact that Americans are emitting six times the global average CO_2 per capita compared to what the rest of the world creates, Patz said, "a big ethical dilemma."

He told the workshop audience how he had the good fortune to show this same cartogram to His Holiness the Dalai Lama at an event in 2011, and the Dalai Lama responded, "If you know pollution kills, your country is not showing much compassion, correct?" When Patz told the Dalai Lama that there was no knowledge about pollution until the 1952 killer London smog event, and little knowledge about climate change until the 1990s, His Holiness responded, "Well, it's 20 years later, and you're still cranking away, polluting and killing people around the world with your energy policies."

Today, Patz said, "we are in a new era of awareness," one marked by the Framework Convention on Climate Change, which was held in Paris in 2015. He considered the conference a turning point, with a record number of heads of state (143) and 183 countries committing to reduce greenhouse gas emissions. The United States committed to reducing its greenhouse gas emissions by 28 percent by 2030; the European Union committed to 40 percent. "These are big commitments," Patz said. However, even with these commitments, the global average surface temperature of the earth is still likely to rise by 3.5°C by 2100 (compared to 4.5°C with no action), Patz explained (see Figure 2-2). Scientists have been warning that it will be necessary to stay below a 2°C warming to avoid catastrophic problems and ecosystem collapse. To decrease global warming to below 2°C, he said, immediate and substantial action is needed.

Cost-Free Climate Change Policy: Opportunities for Health

The way to achieve these immediate and substantial actions, Patz argued, is to not only frame climate change as a public health concern, but also to make the case that policy to combat climate change could be cost neutral. Moreover, with the many public health benefits from climate change policy, especially with respect to non-communicable chronic diseases such as heart disease and cancer, climate change policy could even create a net gain.

Patz identified three sectors where he sees opportunities for a health-in-all-policies approach to combat climate change: (1) the energy sector, (2) food systems (food and agriculture), and (3) transportation and urban planning. He discussed each sector in turn.

The Energy Sector

With respect to potential public health-related policy changes in the energy sector, Patz referred to the World Health Organization estimates of more than 3 million people dying prematurely every year from outdoor air pollution, mostly from urban exposures, and more than 4 million people dying prematurely every year from indoor air pollution, largely from inefficient cookstoves that use biomass (e.g., firewood) and coal. He explained that burning fossil fuels leads not only to the emission of greenhouse gases, but also to the emission of pollutants. In a study on the health co-benefits of a cleaner energy system in the United States, Thompson et al. (2014) predicted that such a system would reduce both ozone and particulate matter ($PM_{2.5}$), with health benefits offsetting the health system's upfront

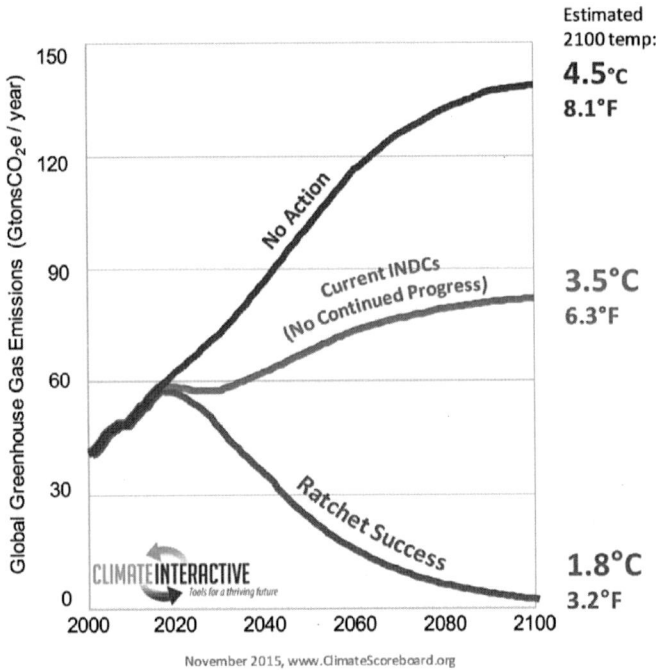

FIGURE 2-2 Three scenarios for the Earth's temperature show the importance of applying a health frame.
NOTE: Predicted change in the global average surface temperature of the earth with no action taken to reduce greenhouse gas emissions (top), action based on current Intended Nationally Determined Contributions (INDCs) (middle), and more immediate and substantial actions than have been promised (bottom).
SOURCES: Patz presentation, March 13, 2017; reprinted with permission from Climate Interactive. Visit https://www.climateinteractive.org/programs/scoreboard (accessed August 1, 2017) for updated figure.

investment by 26 to 1,050 percent. In other words, Patz explained, the health benefits could be up to 10 times greater than the economic cost of investments in green technology.

Policy makers need to understand these health co-benefits of green technology, Patz urged. While it may take more than $30 to remove one ton of CO_2 from the atmosphere by investing in, for example, solar or wind technology, that is only half the equation. Removal of a ton of CO_2 from the atmosphere also removes a lot of $PM_{2.5}$ and other pollutants such that the health benefits could average $200 for every ton of CO_2 removed (West et al., 2013). Thus, there are huge gains to be made by investing in cleaner energy, particularly in highly polluted countries. In East Asia, for example, the benefit could be much greater than $200 per ton, Patz suggested. Moreover, the cost of cleaner energy will likely decrease in the future. Over the past 40 years, solar energy has dropped in price by 99 percent.

"China gets this," Patz said, referring not only to the Chinese president's support for the Paris Agreement, but also the fact that China is the number one producer of solar panels. Additionally, while China does have a coal-fired power plant problem, it recently decided not to move forward with plans for another 80–90 coal-fired power plants, according to Patz, who added, "[e]ven if the United States backs out, China is going to be moving forward."[3]

Food Systems

The food and agriculture sector is another area where environmental policies can have significant health co-benefits. Patz showed a photograph from the 2014 People's Climate March in NYC, with a sign on a large inflatable cow stating, "I am full of greenhouse gas. Do you have a 'steak' in it?" He remarked that many people recognize that eating lower on the food chain is better for both the environment and our health. In a comparison of high meat, low meat, fish, vegetarian, and vegan dieters in the United Kingdom, Scarborough et al. (2014) found that a high meat diet had the highest mean dietary greenhouse gas emissions (i.e., in kilograms of CO_2 equivalents per day, so the amount of energy required to produce one kilogram of protein), followed by a fish diet, then the vegetarian diet, then vegan diet. In another UK study, Westhoek (2014) reported that not only would reducing meat consumption by half cut greenhouse gas emissions by 25 to 40 percent, but it would also reduce saturated fat in the diet by 40 percent, thereby increasing cardiovascular health.

Transportation and Urban Planning

In a 2016 study published in *The Lancet*, NCD Risk Factor Collaboration (NCD-RIsC, 2016) members reported on the increasing numbers of obese and severely obese people around the world by region. Patz asserted that while this unfortunate global trend that affects both men and women is partly food related, it is also related to how cities are designed. Pucher et al. (2010) reported that U.S. cities with the highest rates of walking and cycling to work have 20 percent lower obesity rates and 23 percent lower

---

[3] On June 1, 2017, President Donald J. Trump announced that the United States would withdraw from the Paris climate accord.

diabetes rates, compared to U.S. cities with the lowest rates of walking and cycling. "It is high time that we design cities for people, rather than for motorized vehicles," Patz said.

Patz quoted acting Surgeon General Boris Lushniak, who said at the 2014 Climate Summit, "America does not have a health care system. We have a 'sick-care' system. A health care system extends far beyond medical centers and includes safe routes to school, clean air and water, and flourishing healthy communities." He then cited several epidemiological studies showing that designing communities to promote physical fitness could have huge health gains, including reductions in cardiovascular disease (Hamer and Chida, 2008), diabetes (Jeon et al., 2007), dementia (Hamer and Chida, 2009), depression (Woodcock et al., 2009), colon cancer (Harris et al., 2009), and breast cancer (Monninkhof et al., 2007) (see Figure 2-3). In addition, Patz described how he and colleagues are currently modeling the health effects of an increase in mean walking time, with preliminary data suggesting that increasing people's walking time by just 10 minutes per week, or 2 minutes per work day, would save the state of Wisconsin $30 million in health care and absenteeism costs.

Regarding the current administration's discussion of a trillion-dollar investment in infrastructure, Patz said, "What a golden opportunity for public health," adding that infrastructure is not just highways and bridges, but also bike trails and green communities.

Pivot to Local Leadership

Patz commented that the new administration's views on climate change are different from those of the previous administration. "We really need to start thinking about pivoting the local leadership," he urged. Opportunities for local leadership extend across state public health departments, the health care system, and nongovernmental organizations. Patz also mentioned the C40 city leaders,[4] who have insisted that they "will not be slowed down" if the new administration does not support climate change policy: CDC's Building Resilience Against Climate Effects (BRACE) program (the summary of Paul Biedrzycki's presentation in Chapter 4 includes a discussion of New Hampshire's use of the BRACE framework); Health Care Without Harm (an organization working to transform health care for environmental health and justice); Kaiser Permanente's multifaceted sustainability efforts (see the summary of Kathy Gerwig's presentation in Chapter 6); the pioneering work by Gundersen Health System to invest in renewable energy in La Crosse, Wisconsin (see the summary of Jeff Thompson's presentation

---

[4] C40 is a network of megacities around the world committed to addressing climate change.

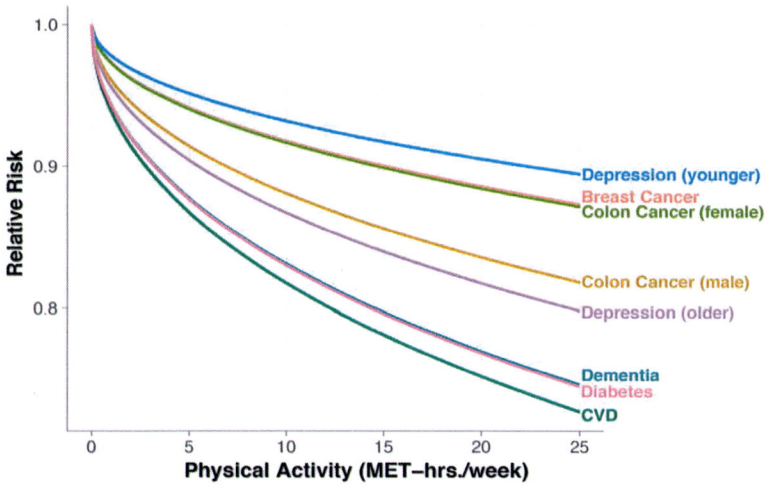

FIGURE 2-3 Effects of physical activity on relative risk for several health outcomes.
NOTES: The relationship between physical activity and relative risk for several
health outcomes, based on epidemiological studies. CVD = cardiovascular disease;
MET = metabolic equivalent.
SOURCES: Patz presentation, March 13, 2017, developed from the following
references: Hamer and Chida, 2008, 2009; Harris et al., 2009; Jeon et al., 2007;
Monninkhof et al., 2007; Woodcock et al., 2009; reprinted with permission from
Patz.

in Chapter 4); the Climate and Health Alliance; and the Medical Society
Consortium on Climate and Health.

In conclusion, Patz emphasized that addressing the global climate crisis
through a low-carbon economy, especially across the energy sector, with
the food system, and in transportation and urban planning, can help make
people healthier and save money. "Doing something urgently about the
global climate crisis," he said, "could be the largest public health opportunity we've had in a very long time."

DISCUSSION

Following his presentation, Patz answered questions from the audience
touching on multiple themes.

Health Professionals and Urban Planners

An audience member, Kelly Dennings, asked whether the public health community is reaching out to urban planners and how this relationship is being fostered. Patz replied that when health professionals show up at urban planning meetings, they are welcomed with open arms. Urban planners appreciate the input. He urged more engagement on the part of the health profession, particularly given that there is receptivity at the other end. As an example, he mentioned the Sustainable City Year Program, which partners universities with cities for service-based learning, and some of these partnerships involve health impact assessments. In fact, one of Patz's classes is currently (at the time of this workshop) conducting a health impact assessment for a nearby city as part of this program. Additionally, he mentioned that the Robert Wood Johnson Foundation has prioritized projects on exercise and livable communities.

Infrastructure Policy: Adaptation and Mitigation

Lynn Goldman mentioned having co-authored a National Academy of Medicine Perspectives paper, *Advancing the Health of Communities and Populations: A Vital Direction for Health and Health Care* (Goldman et al., 2016), which addresses infrastructure and health-in-all-policies around infrastructure. She commented on the need to consider both how to train people going into public health and how to create opportunities for communities to engage public health in decision making around infrastructure. In addition, Goldman remarked that Patz had mentioned walkability and bikeability, but there are many other components of infrastructure as well. For example, too often, when roads are addressed, no one pays attention to the things that run under those roads, namely, drinking water and liquid waste. Referring to a recent National Academies of Sciences, Engineering, and Medicine report attributing many severe weather events to climate change (NASEM, 2016), Goldman pointed out that infrastructure also encompasses aspects that are important for protection from severe weather events. She asked Patz to reflect on steps that need to be taken to prepare communities for these events, while at the same time making communities more aware that steps also need to be taken "on the longer haul" to prevent such events from becoming even more serious and more frequent. She observed that science is increasingly attributing many severe weather events to climate change, yet, she said, "We're still very cautious, I think, as a community about being clear to the public about that."

Patz replied, "We should not be thinking either/or," meaning either adapt or mitigate. In his view, communities need to take a two-pronged approach. They need to be ready for what is already happening. He agreed with Goldman that there is a complexity to adaptation, for example, with respect to roads being built for walking, biking, and motorized vehicle safety. He called attention to the ecological aspects of infrastructure as well. For example, building a sea wall to hold the ocean back as an adaptation to sea level rise can kill an estuary or mangrove, thereby destroying fisheries in the region. "So when you think about adaptation, first, do no harm," he said. But then, think more broadly about potentially damaging unintended consequences. He called for a more interdisciplinary adaptation approach and greater engagement across disciplines. In addition to being prepared, at the same time, he said, "We also need to go upstream and be mitigating the root problem to reduce fossil fuel emissions."

Air Pollution

Phyllis Meadows of the Kresge Foundation asked Patz whether the fact that indoor air quality is contributing to more deaths worldwide compared with outdoor air quality threatens "the narrative" and whether indoor and outdoor air quality issues should be discussed in tandem to engage a broader audience. Patz replied that, depending on where in the world you live, the intervention and health benefit will be different. In poor countries where people cook with coal and wood and where there is a lot of $PM_{2.5}$ exposure indoors, using cleaner cooking fuels is important. In these areas, especially in East Asia, indoor air pollution is an enormous problem. But in the United States, for example, there is more to gain by addressing outdoor air pollution.

Tree Canopy Spread as a Mitigation Strategy

Terry Allan from the Cuyahoga County Board of Health, Ohio, re-marked that there has been a lot of recent talk in urban planning around tree canopy spread as a mitigation strategy. He asked Patz to comment on the importance of and opportunities for tree canopy spread from a public health perspective. Patz replied that, given that most people live in an urban environment, it is extremely important to be smart about prevention in this environment. During his talk he had mentioned the role of walkability and bikeability in increased exercise. The greening of cities, he said, not just through trees, but also, for example, by growing green rooftops, plays a very important role in reversing the urban heat island effect seen in built-up areas that experience higher temperatures than surrounding rural or less developed areas due to pavement and roofs that limit vegetation and moisture.[5] He added that trees also reduce rainfall runoff and contamination. Finally, he mentioned the mental health benefits of a green canopy and a Wisconsin study showing that the extent of green cover canopy in an urban environment has as much of an effect on mental health as moving or divorce. One caveat about trees, he noted, is that some trees emit more volatile organic compounds, which are a precursor to ozone smog pollution, and others also give off more pollen, underscoring the importance of carefully selecting tree species to plant in cities.

Returns on Investing in Green Energy

"The return on investment data that you show is almost shocking," David Kindig of the University of Wisconsin remarked. He wondered whether any work being done with social impact bonds[6] has shown similar returns, adding that he was unaware of a similar level of return for bonds in the areas of early childhood or criminal justice. Patz replied, "It's no surprise, really, the $30 versus $200." He recalled that when EPA's Clean Air Act required benefit-cost assessments, it was discovered that every $1 investment in clean air returns $30 or more. These numbers, in his opinion, are critical, especially with the new administration beginning to roll back regulations. He urged a greater understanding of the value of these regulations. He noted that EPA has announced an assessment of climate change policies and suspected that such an assessment will likely examine only the energy cost side of the equation (i.e., the $30 investment), without considering the health benefit side (i.e., the $200 benefit). This is impetus, in his opinion, to "get the public health issue on the table."

Kindig also mentioned the Internal Revenue Service community benefit requirement for nonprofit hospitals and recent discussions about moving that benefit in a direction more explicitly oriented toward improving population health. He asked if Patz had seen this type of investment as an important part of any nonprofit hospital portfolio. Patz mentioned Jeff Thompson's work with the Gundersen Health System in La Crosse, Wisconsin, where they are reinvesting into the community and seeing great gains in health promotion and disease prevention (Thompson's presentation is summarized in Chapter 4).

---

[5] The Environmental Protection Agency provides information about urban heat islands on its website; see https://www.epa.gov/heat-islands/learn-about-heat-islands (accessed May 9, 2017).

[6] Social impact bonds, also known as pay for success financing, "allow philanthropic funders and private investors to pool capital for social programs, with the loans repaid by the government only if the funded initiative achieves agreed-on results" (IOM, 2015, p. 35).

Vulnerable Populations: Children, Elderly, and the Poor

Debbie Chang of Nemours asked Patz if children have been included in any of the return on investment studies and if there have been any targeted approaches for addressing health equity issues. Patz replied that, according to the World Health Organization, 88 percent of climate change impacts affect children. Diarrheal disease, malnutrition, and malaria, for example, are all very climate sensitive and strongly affect children. In the United States, children are particularly vulnerable to allergens (i.e., asthma). Older adults are also vulnerable, particularly with respect to heat waves. Finally, Patz reminded the workshop audience that it was people who could not afford to leave town that suffered the most from Hurricane Katrina.

Environmental Refugees

Sanne Magnan of the University of Minnesota asked Patz to speak more to the issue of environmental refugees. Patz replied that the fact that the Syrian Civil War was preceded by the worst drought in instrumental record makes it very difficult to tease out the impact of climate change. The extreme environmental conditions of climate change—not just drought, but also sea-level rise—push people around a lot, he said. That destabilization alone, with people being forced to move and resettle and either having no immunity to new diseases or bringing diseases with them, creates a huge public health burden. He urged that, even though it is difficult to quantify the effect on refugees, this destabilization is something to keep in mind.

Maureen Lichtveld of Emory University added that there are environmental refugees in the United States as well. She referred to the communities facing displacement due to the retreat of the bayous and loss of land on the Gulf Coast.

Engaging Leaders at the National and Global Levels

Benjamin Miller, an audience member from RAND Corporation, agreed with Patz's suggestion to engage local leaders, but said, "This is also a national and global issue." He asked Patz how to engage those leaders while also engaging local leaders. Acknowledging that there will always be uncertainty, Patz replied, "We absolutely need to be better at communicating the truths that we know, or at least the best of science that we know." He expressed discouragement that the head of EPA does not think there is a relationship between CO_2 and climate when, in fact, a great deal is known about that relationship. He mentioned a study showing that 97 percent of climate scientists agree that not only is global warming occurring, but the warming is attributed to human activity, mostly the burning of fossil fuels

(Cook et al., 2016). He urged keeping in the public ear the fact that a huge majority of climate scientists agree that climate change is real. "Pivot to local leadership," he said, "but also keep informing the general public."

How to Talk About Climate Change

Magnan reflected on the skepticism that many policy makers and the general public have toward climate change. She asked Patz, based on his work, what perspectives or stories most resonate with people who are skeptical.

The most important messages to communicate, Patz replied, are "Climate change is real. It's now. It's already happening. It's solvable." Patz clarified that there is no scientific debate about whether climate change is happening, other than a few outliers. The debate, he said, is political. Moreover, it is actionable. He mentioned again that the price of solar energy has dropped. Wind energy has also become competitive. These technologies are not things that we have to wait for; they already exist. "We can do something right now," he said.

In addition to these messages, he suggested not using the term "climate change" when talking to people who do not believe in climate change. Talk instead about extremes in climate variability, water stress, or heat waves. Instead of talking about interventions to mitigate climate change, talk about reducing fossil fuel combustion and the immediate health benefits of doing so (i.e., better food quality, exercise promotion, cleaner air quality).

George Isham reflected on the fact that people without a science background may not be able to evaluate highly technical or scientific descriptions. He suggested that one strategy for improving public understanding is to identify the top 10 things to know in order to understand an issue or topic beyond the political [debate] level. In response, Patz said the first thing even experts need to do is to become expert listeners. "Our engagement has to be much more participatory," he said.

Lichtveld added that, in her opinion, perhaps the highest priority message coming out of this workshop is to increase environmental health literacy. "Whether it is here, whether it's in New Orleans, or across the world," she said, "it is that, that will empower us."

Political Organizing and Partnerships

Patz was asked by an unidentified audience member to expand on ways to communicate science-based evidence in light of the reality that the current federal administration may not be influenced by such evidence. The audience member expressed dislike of the phrase "postfact era," but admitted "that seems to be what we are in." The audience member asked how the federal policy space could be more assertively entered and suggested that perhaps partners, some unusual allies, may play a role. He asked, "Where are the edges? Where are the barriers?"

Patz agreed that finding good partners and coalitions is absolutely necessary. Although scientists need to be better communicators, he agreed that they are not necessarily going to convince people. He mentioned the Climate and Health Alliance, which was represented at this workshop by Linda Rudolph (Rudolph moderated the Session 1 panel discussion, summarized in Chapter 2), and said the most trusted professionals in the United States are nurses (the Climate and Health Alliance represents health care professionals from a range of disciplines, including nurses). "Finding these new coalitions is extremely important," he said.

Additionally, Patz urged visibly engaging with and investing in disadvantaged populations, coal miners in particular. He stressed the importance of job diversification. Otherwise, as we shift away from coal, coal miners will be left behind. He said, "We need to be out there taking care of them and being very proactive about that."

3

Regional Perspectives from the South

Moderated by Linda Rudolph, Panel 1 speakers addressed a range of city and state-wide climate change health scenarios and strategies being implemented and developed in the South, specifically in Kentucky.

Maria Koetter opened the panel with a presentation on findings from two recent city-wide urban heat and tree buffering analyses and steps being taken by the city of Louisville to address climate change and its impact on health.

In the question and answer period at the end of the panel, she emphasized the important role of strong city leadership and an awareness among city leaders, including the mayor, regarding the issues. Next, Halida Hatic and Rachel Holmes provided a joint presentation on a Louisville project, Landscape Audit for Sacred Spaces, that serves as a tool for religious organizations to conduct tree canopy assessments and evaluate other components of their landscape and their relationship with that landscape. The fourth and final panelist, Lisa Abbott, described Kentuckians for the Commonwealth's (KFTC's) work to develop a grassroots People's Plan for the state's energy future. This chapter summarizes these presentations. Key points made by each panelist are listed in Box 3-1.

The open discussion with the audience at the end of the panel covered a range of issues: the role of the health sector in local efforts to address climate change; building political will; engaging the private sector; engaging religious networks; and the concept of "just transition" (i.e., from traditional energy, namely coal). This discussion is summarized at the end of this chapter.

BOX 3-1
Key Points Made by Individual Speakers

- The city of Louisville is relying on results from a recent urban heat analysis and city-wide tree canopy assessment to guide its efforts to manage what has emerged as the most rapidly growing heat island in the country. (Koetter)
- As part of its comprehensive sustainability plan, a range of other climate change-related activities are also in progress in Louisville. These include cool and green roof incentives, a cool paving project, buffering (i.e., against air particulates) and tree planting pilots, and a communication campaign. (Koetter)
- A key to successful partnering, which plays a vital role in Louisville's Landscape Audit for Sacred Space project, is to "check your agenda at the door." (Hatic)
- In addition to mutually beneficial partnerships, the Landscape Audit for Sacred Spaces project has relied on leveraging already existing tools and methodologies, including what The Nature Conservancy has been doing with urban forestry across the country and what GreenFaith has been doing with audit programs. (Holmes)
- Soliciting input from and engaging Kentuckians from across the state and from all sectors was an important piece of the process used to put together Kentuckians for the Commonwealth's People's Plan to facilitate a just transition from a coal-based economy. (Abbott)
- Among several other predicted outcomes expected by 2032, if the People's Plan recommendations are met, CO_2 emissions from the power sector will be reduced by 36 percent, health will be improved, and jobs will be created. (Abbott)

LOUISVILLE METRO GOVERNMENT
OFFICE OF SUSTAINABILITY[1]

Maria Koetter opened the panel by sharing work under way in Louisville's Office of Sustainability. Currently, the office is working through a comprehensive sustainability plan, with a focus on energy efficiency, green infrastructure (i.e., tree canopy, reforestation), and urban heat management.

Urban Heat Analysis: Findings from Louisville

In 2012, Brian Stone, Georgia Institute of Technology, released a study showing that Louisville had the most rapidly growing heat island in the country from 1961 to 2010. This finding, Koetter said, raised a red flag,

[1] This section summarizes information presented by Maria Koetter, LEED AP, Director of Sustainability, Office of Sustainability, Louisville Metro Government, Kentucky.

prompting an application for a grant from Partners for Places. The grant was awarded, allowing the city to hire Stone to conduct an in-depth analysis of Louisville's urban heat island issues and to model strategies to help manage the heat. His analysis also included population vulnerability and heat mortality findings.

In addition to Stone's analysis, the city has also conducted a city-wide tree canopy assessment. Koetter described Louisville as having a consolidated city–county government, with 750,000 people living across approximately 400 square miles. "So our purview is broad," she said. The tree canopy analysis showed that areas with the lowest tree canopy coverage (i.e., downtown, up in the northwest quadrant of Louisville) overlap with areas that have lower socioeconomic status, higher disease mortality rates, and 10-year shorter life expectancies than other areas in the community (see Figure 3-1). Koetter remarked that diabetes and asthma also disproportionately affect this same downtown area. Additionally, people living in areas with lower socioeconomic status bear a disproportionately higher economic burden associated with the changing climate. This is because

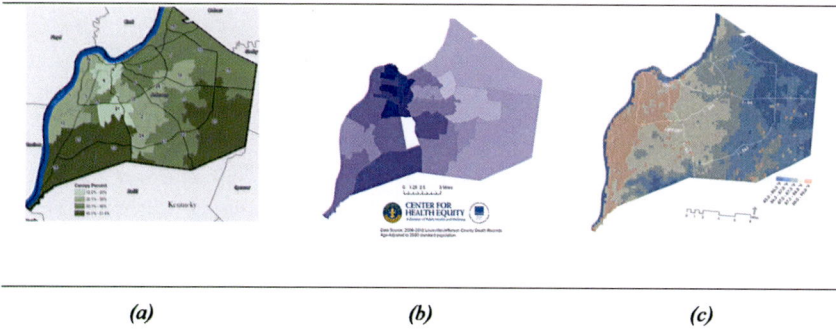

(a) (b) (c)

FIGURE 3-1 Three maps of Louisville showing (a) tree canopy coverage, (b) mortality from heart disease, and (c) average daily high temperature (May to September 2012).
NOTE: Maps illustrate overlap in geographic variation in tree canopy coverage, average daily high temperatures (May to September 2012), and mortality from heart disease across the city of Louisville.
SOURCES: Koetter presentation, March 13, 2017; (a) Louisville Office of Sustainability, 2016; (b) created using 2006–2010 Louisville/Jefferson County Death Records Age-Adjusted to 2000 standard population; reprinted with permission from the Louisville Center for Health Equity director, Dr. Kelly Pryor; Louisville Center for Health Equity, 2014; (c) Louisville Office of Sustainability, 2015.

these neighborhoods, Koetter explained, tend to have older housing stock that is less insulated, and thus, residents tend to spend more on utilities for heating and cooling.

Stone's urban heat analysis showed that areas of the city that feel the heat the most (i.e., based on air temperature readings, which Koetter explained is where people actually feel the heat, as opposed to surface temperature readings) are the same areas that were shown in the tree canopy assessment to have lower tree canopy coverage, worse health outcomes, and shorter life expectancies.

The urban heat management strategies that Stone modeled in his analysis covered greening methods, cooling methods, and all methods combined. His results showed that greening certain areas of the community can reduce heat by more than 2°F. But the amount of greening required to do this, Koetter said, "is not a small task." Specifically, it would require planting 450,000 trees and installing 730 10,000-square-foot green roofs. Achieving similar benefits from cooling methods, rather than greening methods, would require installing only 23,000 roofs that are 10,000 square feet, but paving 54 miles of road, surface area, parking lots, and other areas with cooling materials. The findings that most surprised and pleased Stone, according to Koetter, were that the effects of all methods combined were far more than additive. For Koetter, these results highlight the need to consider multiple strategies to manage urban heat.

Planting Trees: Effects on Air Pollution and Health

Koetter continued to describe another recently completed Office of Sustainability project, Green for Good, which was funded through a Partners for Place grant. The goal was to install a densely vegetated buffer between a population of people and a heavily trafficked road, and evaluate particulate matter (i.e., in the air) and conduct biomarker sampling (i.e., in volunteers) both before and after installation of the buffer.

After assessing more than 20 properties, the city chose an elementary school as the study site because it was known to have high-traffic pollution and because it had space for planting the buffer. A densely vegetated buffer was planted between the street and the school population, with part of the school yard left unbuffered to serve as a control. In addition to air monitoring, the city also conducted both before and after blood and urine biomarker sampling of 80 volunteer students and teachers at the school. All of the pre-intervention air monitoring and biomarker sampling occurred in September, the planting of the buffer (100 trees) occurred in October, and the post-intervention measurements were collected in November.

The air monitoring showed a 60 percent reduction in particulate matter behind the buffer, compared with the front of the street. The biomarker sampling indicated that epithelial progenitor cells increased in number, and immune cell numbers decreased. According to Koetter, the biomarker findings are an indication that the buffer worked, and the city of Louisville is partnering with the University of Louisville to develop a preliminary report, with a full study of the results from Green for Good to be published in a peer-review journal.

Next Steps for Louisville

Koetter listed several next steps for Louisville, starting with the city having been recently selected for the final round of the 100 Resilient Cities program, which is funded by the Rockefeller Foundation. According to Koetter, as part of this program, the city's chief resilience officer will be addressing climate change issues related to health and the environment.

Additionally, Louisville's Office of Sustainability recently released its Cool Roof Rebate Program for residential and commercial projects, providing $1 per square foot and ranging from $2,000 to $10,000 per project. The goal of the program, Koetter explained, is to inspire and incentivize installation of Energy Star cool roof shingles, which help to not only manage the heat island effect across the city, but also to reduce energy bills. Sixty percent of that funding has been set aside for neighborhoods that feel the most heat (i.e., neighborhoods in the northwest side of the city). The city also offers an incentive for green roofs.

In addition to these cool and green roof incentives, the Office of Sustainability is conducting a cool paving pilot project across the city. Koetter remarked that there are many materials that are cooler or more highly reflective than concrete or asphalt. The pilot project aims to set an example for what materials can be used and how such materials can be laid successfully to withstand four-season weather.

Another pilot project currently under way is the planting of nine trees in a diamond configuration in the center of the parking lot across from the city hall. The area where the trees were planted was topped with porous paving materials so that cars can drive over it without damaging the trees. It is a small-scale project, Koetter commented, but one that will, with success, demonstrate to private property owners what they can do to green their parking lots without losing any parking spots. With respect to the tree canopy at large, the city-wide goal is to achieve 45 percent coverage. Coverage is currently at 37 percent. The focus is on parts of the community where people live. "We really need to inspire citizens, because the most available property is on private lands," Koetter said.

In addition to roof and paving projects, as well as tree canopy work, the city has been focusing on messaging. For example, the city advertises on city buses to not only promote awareness of extreme heat events, but for all greening, cooling, and other energy conservation programs across the city (using the social media hashtag #cool502, in reference to Louisville's area code).

Finally, the city is in the process of updating its comprehensive plan, which Koetter identified as a great opportunity to integrate not just health, but also sustainability, into all policies. In addition, the city is currently conducting a greenhouse gas inventory and will be setting a target reduction goal as part of that process. There is also some interest in predicting the community health effects of the city's carbon reduction efforts. A final next step, she said, will be to enlist the help of a biostatistician to integrate all of the various types of data that have been collected related to the city's greening, air quality, and energy efficiency programs, and to health outcomes.

FAITH IN LANDSCAPE STEWARDSHIP[2]

In a joint presentation, Halida Hatic and Rachel Holmes described a project in Louisville that was co-created by the Center for Interfaith Relations (represented by Hatic), a Louisville-based nonprofit organization that works to promote interfaith understanding, cooperation, and action; The Nature Conservancy (represented by Rachel Holmes); and GreenFaith.

In her opening remarks, Hatic described the project, Landscape Audit for Sacred Spaces, as a methodology, or protocol, that was jointly developed by the partners listed above. The intent of the project, she said, is to help communities of faith learn about the ecological benefits of their properties and identify achievable goals to improve their landscapes for both people and nature. Ultimately, the intention is also to help solve some of Louisville's more challenging problems.

Hatic mentioned that she would be speaking first on the important role that strong partnerships serve in Landscape Audit for Sacred Spaces and how cultivation of a shared vision for the project is what led to these strong partnerships. She described these partnerships as mutually beneficial, organic, honest, creative, and most importantly, in her opinion, successful. Then, Holmes would discuss the methodology and tools being used in the project. Hatic commented that the project did not have to "reinvent the wheel" with respect to the tools employed, but they did have to connect and listen to put the tools together in a way that would support a diverse, multifaith audience.

---

[2] This section summarizes information presented by Halida Hatic, director of Community Relations and Development at the Center for Interfaith Relations in Louisville, Kentucky, and Rachel Holmes, conservation coordinator for Healthy Trees, Healthy Cities, The Nature Conservancy.

Landscape Audit for Sacred Spaces:
The Importance of Mutually Beneficial Partnerships

Hatic referred to Koetter's description of Louisville's many challenges, including its poor air quality and the fact that it has one of the fastest growing urban heat islands in the nation. She remarked that, while there is much work to be done, Louisville has also made huge strides in turning these challenges into opportunities to improve the health and well-being of its citizens. Louisville is a leader in the compassionate cities movement and has earned the title of Model International Compassionate City 5 years running. Earning this title, Hatic said, takes a lot of political will and a lot of courage on the part of city leadership. Hatic remarked that in addition to being in the 100 Resilient Cities network, Louisville has offered itself as an "urban laboratory" for exploring problems and possibilities and for sharing its wisdom and experiences with the world.

Finally, for a number of different reasons, according to Hatic, Louisville has a religiously and culturally diverse community with a rich history of celebrating, promoting, and supporting interfaith collaboration and action. For example, the Center for Interfaith Relations hosts an annual event called the Festival of Faiths, which is now in its 22nd year. Through this collaboration and action, there has been much dialogue in the faith community about the issues being addressed at this workshop. Hatic emphasized, as well, that not only are people and communities of faith emerging at the local level, but their voices are emerging globally and as leading voices in the collective environmental consciousness. Framing climate change as a moral issue is not a new issue, in her opinion, and it is one that she predicts will be heard over and over again in the future through voices like those of His Holiness the Dalai Lama and Pope Francis.

The genesis of the Landscape Audit for Sacred Spaces, Hatic continued, was four people with shared values and the desire to collaborate who were sitting around a table with what Hatic described as "truly noble intentions." "We had no idea what we were going to come up with," she said, "but we knew we wanted to do something meaningful together." At the time, the Center had been partnering with the Cathedral of the Assumption, the Catholic cathedral in Louisville, on an energy audit. In conversations at the table, with The Nature Conservancy present, participants realized that there was nothing equivalent to an energy audit that could be used to assist faith communities with assessing the health and well-being of their living landscapes. Thus, the idea to co-create the Landscape Audit for Sacred Spaces methodology was born.

Of the three partners, the Center for the Interfaith Relations was able to bring the people and "street cred," Hatic said, while The Nature

Conservancy brought the science and technology, and GreenFaith, the amplification.

The Center hosted three in-person meetings, which included feedback sessions and trainings, over the course of the first several months. Volunteers were then sent off to implement the pilot methodology, with support provided by the partners. "It was a pretty simple process," Hatic said, "but had huge results."

Four institutions participated in the pilot: Center for Interfaith Relations, St. Xavier Catholic High School, River Road Mosque/Louisville Islamic Center, and Christ Church Cathedral (Episcopal). These organizations, Hatic explained, covered a diversity of locations, in terms of urban versus suburban environments, as well as a diversity of age participants and faith traditions.

In conclusion, Hatic reiterated the importance of checking your agenda at the door for a partnership to be successful. The relationships need to grow organically, she emphasized. She also emphasized the importance of being flexible, starting at a base of shared values, gaining mutual benefit, and having a shared need.

Landscape Audit for Sacred Spaces: Usable Tools and Support

Holmes began by expressing what a wonderful privilege it was to be at a workshop where the public health officials are in the majority, not the minority, and it is the conservationists, like herself, who "are representing" their field. She reiterated what Hatic had said about the reality that the Landscape Audit for Sacred Spaces partners did not need to reinvent the wheel when developing their methodology. Rather, the methodology and tools came out of work that The Nature Conservancy had been doing with urban forestry across the country and that GreenFaith had been doing with audit programs.

A key to this work, Holmes stated, is to not stop at the data. The key is to analyze and plan for action. "It is not enough just to know what you have out there," she said. It is more important, she went on, that you know how to manage what you have. She commented on the "beauty" of the process when, for example, someone using a clipboard or app to collect data begins to see not just a wooded area behind a basketball court, but that the wooded area is actually a critical wetland that supports wildlife.

While doing the best they could to leverage tools and methodologies that they knew already existed, the partners also wanted to develop an audit and protocol that were as accessible to as many people as possible. The intention, Holmes explained, was for the audit, when it goes national, to be accessible to anyone, regardless of what funding, skill level, and knowledge base exists in their community.

In sum, the guiding principles for the audit were no cost for participation, simple equipment/protocols, well-supported methods (e.g., videos and fact sheets are available for people who need instruction), accessibility for differently abled people (e.g., some people may not be comfortable outdoors and would rather participate by being indoors, uploading data), and community connectivity (i.e., ensuring that what people are reading about their landscape or gaining from the audit is directly applicable to partners in the community).

The audit has four components: (1) tree canopy assessment, (2) landscape mapping, (3) grounds management, and (4) worship/community. As an urban forester, Holmes expressed excitement that, during the initial focus group sessions, people prioritized tree canopy assessment. She noted that, in addition to the information presented by Jonathan Patz, another important role for trees is the absorption of particulates. The tree canopy assessment component of the audit encompasses not just identifying what trees are present, but also where they are situated and, more importantly, how healthy they are, including whether they are being affected by any insects or diseases that are known threats to the urban tree canopy (e.g., the emerald ash borer).

Holmes also highlighted the worship/community component of the audit, which, she explained, involves using a basic questionnaire to ask if concepts of nature and stewardship are being integrated into liturgical practices whenever possible. Or, instead, do these concepts only surface around Earth Day? There may be other opportunities for talking about what it means not just to love nature, but to take care of nature.

Currently, the project partners are in the process of taking the information and feedback received from the pilot and making updates and diversifying opportunities. Holmes said they are envisioning a fall 2017 or spring 2018 national launch.

Holmes reiterated Hatic's message to "check your agenda at the door." The Nature Conservancy, she said, did not enter the door and say, "Here's a tool. Go ahead and use it." Rather, they asked what information would be beneficial to the community. She quoted participants from two partner organizations, one from the Louisville Islamic Center's Brother Sikhander Chowhan who said, "If you don't care for your place of worship . . . that's in a way disrespectful to our Creator." The second quote was from the Very Reverend Joan Pritcher of Christ Church Cathedral (Episcopal), who said that opportunities like the audit "appeal to people we aren't as successful reaching through traditional faith community activities."

In closing, Holmes contrasted people's motivations from a faith perspective (redemption, forgiveness, stewardship, ritual, preaching, enlightenment, praise, prayer) versus those from a conservation perspective (restoration, mitigation, stewardship, best practices, teaching, analysis, celebration) and asserted that, regardless of whether you are drawn to conservation because you are motivated by a belief or, instead, by ethical humanism, "it's time to join the discourse." She was aware that many people feel uncomfortable using specific (e.g., religious) words in conversations about landscape stewardship. It is okay, she said, to say, for example, "I pray," adding about herself, "I'm a conservationist. I'm in the pew on Sunday, but I'm out in the field on Monday." She urged participants to show leadership in bridging these motivation and language gaps.

EMPOWER KENTUCKY[3]

Empower Kentucky is an effort to shape a people's energy plan for the state of Kentucky, Lisa Abbott began. The project was launched immediately after the Environmental Protection Agency (EPA) announced its Clean Power Plan, which state political leaders from both sides of the aisle opposed, according to Abbott, with both candidates for governor vowing to not comply if elected. Thus, it was thought that perhaps a grassroots social organization could take on the challenge and, through a meaningful public process, develop a plan with outcomes that are better for health, better for jobs, and better for average [utility] bills; advance a just transition for affected workers; advance racial and economic justice; and comply with the Clean Power Plan. "So that's what we set out to do—no small task," Abbott said.

Abbott described Kentuckians for the Commonwealth, the parent organization for Empower Kentucky, as a 35-year-old grassroots organizing group working for a better quality of life for all Kentuckians and for a vibrant and healthy democracy. KFTC believes, Abbott continued, that right now in Kentucky, as well as elsewhere around the world, a real opportunity exists to shape a just transition to a clean energy economy. The year 2015 was the first time that coal's share of Kentucky's electrical energy generation fell below 90 percent, to 87 percent. Although the energy and political landscape supported by coal have been in place for decades, Abbott remarked that it is changing rapidly due to many factors, including that the state's aging fleet of coal plants is retiring. She said, "All of our energy eggs have been in one basket for decades. . . . We have a choice about . . . the next economy . . . [and] the next energy system."

The stated goals of Empower Kentucky are to generate political will for a just transition to a clean energy future; meaningfully engage thousands of Kentuckians; develop an environmental justice analysis for Kentucky; and develop a people's plan for Kentucky's energy future. Abbott clarified that

---

[3] This section summarizes information presented by Lisa Abbott, organizer, Empower Kentucky, Kentuckians for the Commonwealth, London, Kentucky.

she would be presenting preliminary results and that the results and report would be released in mid-April 2017.[4]

A Grassroots Effort to Change Kentucky's Energy Future: The Importance of Process

"In times of transition, process really matters," Abbott emphasized. It matters, she continued, that people have their voices and stories heard and that there is input and discussion. Thus, among other parts of the Empower Kentucky public process, KFTC organized community conversations in all six congressional districts of Kentucky in spring 2016. The conversations were 3-hour programs that included locally sourced meals and facilitated table discussion.

The community conversations began with asking participants to spend three minutes each telling the story of relationship with Kentucky's energy system. Abbott described the "intimate" relationship that Kentuckians have with the state's energy system—one stemming not just from the work they have done in the mines, but also the asthma they have experienced, the trouble they have had paying their bills, and the host of other ways people's lives have been deeply connected to the state's energy system. If KFTC were to have started the community conversations by talking about abstractions, they would have missed the opportunity to learn about these connections. Following the sharing of stories, participants were asked to address some key specific questions: What is your vision for Kentucky's energy future? What do you think that will take? How can we ensure that all Kentuckians will benefit and that no one is left behind as we transition to a clean energy economy? Every seat at every table at every community conversation was filled, Abbott remarked. The seats were free, she noted, but residents had to reserve their seats ahead of time.

In addition to these community conversations, KFTC also conducted an online survey and house meetings. A total of 750 people took part in the community conversations, and 1,200 people from across the state provided input into the plan. Abbott described the effort as "very intentional" in encouraging input from a broad and diverse range of communities.

Among other results, when asked about their vision for Kentucky's energy future, residents expressed a very strong interest in renewable energy, especially solar; health; job creation; and the opportunity for a just transition for workers from the fossil fuel industry.

In addition to seeking public input, Empower Kentucky also undertook its own economic justice analysis. Abbott explained that when it issued the

---

[4] See http://www.empowerkentucky.org/wp-content/uploads/2017/04/Empower-Exec-Summary-Report-final-1.pdf (accessed January 5, 2018).

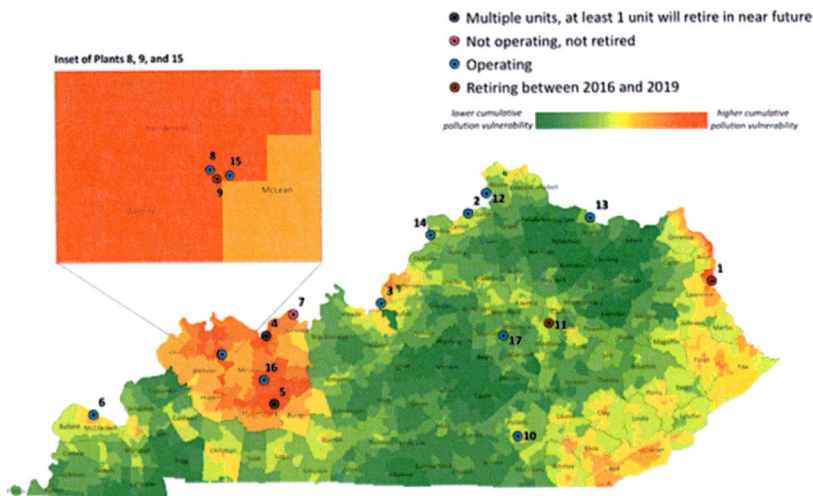

FIGURE 3-2 Cumulative pollution overlay.
NOTE: Operating and retiring coal plants in Kentucky, overlaid by cumulative pollution (i.e., as measured by about a dozen indicators of pollution, not all of which are related to the energy system [e.g., lead exposure]).
SOURCE: Abbott presentation, March 13, 2017. Reprinted with permission from Kentuckians For The Commonwealth, 2017.

Clean Power Plan, the EPA required that states have meaningful public engagement, but did not define that engagement, and they recommended that states do their own environmental justice analyses. Empower Kentucky's environmental justice team is composed of about a dozen volunteers from around the state, led by a recent graduate of University of Kentucky. Abbott showed the map of coal plants in Kentucky with an overlay showing cumulative pollution (i.e., as measured by about a dozen indicators of pollution) (see Figure 3-2) as an example of the findings that have emerged from the environmental justice analysis. Abbott pointed out that the two areas where coal mining takes place (one in the east, the other in the west) also have the greatest cumulative pollution, and people living in the western part of the state are doubly exposed to both extraction and burning.

The People's Plan:
Framework, Recommendations, and Predicted Outcomes

Out of the public process, combined with results from the environmental justice analysis, Empower Kentucky is developing a People's Plan (released several weeks after the workshop, in April 2017).[5] The plan's framework has seven components: (1) accelerate energy efficiency and renewable energy across the state's economy; (2) create jobs and support a just transition; (3) prioritize health and equity; (4) support local solutions; (5) invest fully and fairly in the energy transition; (6) meet our obligations to the climate; and (7) engage everyone to change other essential systems. Regarding the last component, Abbott explained, it is not just the energy system that people across the state want to change. They also want to change the food system, transportation, and other essential systems.

The plan is long, Abbott said, with recommendations encompassed within each of the seven components above. She described a few key components. First, the plan recommends that Kentucky significantly ramp up its efforts in energy efficiency in order to save, over a 15-year period (i.e., by 2032), 15 percent[6] of what would otherwise be consumed, with a strong emphasis on achieving 18 percent savings through programs that directly benefit low-income households. Second, the plan recommends that Kentucky meet a 25 percent renewable energy goal, with at least 1 percent coming from distributed solar. Third, the plan recommends putting a price on carbon. This recommendation was made for two reasons: first, to generate revenue to invest in a just transition for coal workers and coal communities, and second, because an analysis conducted by consultants showed that even if Kentucky were to reduce the amount of "dirty energy" being used by Kentuckians, neighboring states would continue to buy Kentucky's coal-fired power. In other words, Abbott said, "We would become the designated smoking area." Attaching a price to carbon will make their coal power more expensive and less competitive in the regional market.

Empower Kentucky participants examined many possible prices, Abbott explained, including the Obama administration estimate that the true cost of carbon is just under $40/ton. They chose to recommend a low $1–$3/ton, because it was determined that a higher price would accelerate the rush to natural gas and that this lower price was sufficient to reduce emissions and achieve the 15-year goal for implementing more renewable energy and for more energy efficiency. A fourth key recommendation of the plan is to not allow biomass to count as "low carbon" or "carbon neutral."

The preliminary predicted outcomes,[7] by 2032, include

---

[5] The People's Plan was released in April 2017 and is available at http://www.empowerkentucky.org/wp-content/uploads/2017/04/Empower-Exec-Summary-Report-final-1.pdf (accessed May 26, 2017).

[6] The numbers shared were preliminary. In the Final Empower Kentucky Plan this percentage is 17.

[7] See http://www.empowerkentucky.org/wp-content/uploads/2017/04/Empower-Exec-Summary-Report-final-1.pdf (accessed January 5, 2018).

- More jobs than business as usual (46,300 more job years);
- Improved health (from lower sulfur dioxide and nitrous oxide emissions);
- Lower average bills ($13 less than business as usual);[8]
- The generation of nearly $400 million to be reinvested in a just transition for coal workers and coal communities;
- A 36 percent reduction in CO_2 emissions from the power sector (the Clean Power Plan's goal for Kentucky was 31 percent, so this slightly exceeds that goal, Abbott noted);[9]
- An $11 billion investment in energy efficiency; and
- A priority on low-income and industrial energy savings.

Some additional predicted outcomes are that fewer gas plants will be built, compared with business as usual; the same number of coal plants will be retired, compared with business as usual, but with the existing fleet being used less than it is now (i.e., the direction the state is heading, Abbott explained, is that old coal plants are being retired regardless of whether "we do good things"); [production facilities for] another gigawatt of solar and another 600 megawatts of wind will be built; and less electricity will be imported, compared to business as usual.

The People's Plan as a Foundation for Future Work

Abbott acknowledged the very difficult political environment into which these outcomes will be released, both at the national level and within Kentucky. She encouraged viewing this plan as part of creating a broader political will to "do good things" and produce better outcomes for Kentuckians in terms of jobs, health, bills, and the climate. The intention is to use the People's Plan as a foundation for alliances, collaboration, and campaigning across the state.

"We see the biggest opportunity at a very local level," Abbott said. She mentioned, as examples, calling on utilities in Louisville to achieve greater energy efficiency, and other local campaigns to get more all-electric buses on the streets.

Already, Abbott observed, the Empower Kentucky process has strengthened the network and political will in Kentucky to resist moving in a direction that could be detrimental to health and well-being, and to continue to work for policy change "where we can." She mentioned a very powerful

---

[8] In the final Empower KY Plan, the average residential bill would be 10 percent less than under the business as usual scenario and the percentage was used rather than $13.

[9] In the final plan, CO_2 reduction is 40 percent by 2032 (from 2012), not 36 percent. All other numbers remained as reported at the workshop.

senator in the state legislature having recently introduced a bill that would have decimated Kentucky's homegrown solar industry, which Abbott said is small, but growing. However, the senator was so shocked by public outcry that he removed the bill. She concluded, "This project, I think, can produce some of those kinds of results."

DISCUSSION

Following Abbott's talk, panel moderator Rudolph commented on the "inspiring" nature of the panelists' talks in terms of what can be done through local action to address these problems in a positive and constructive way. Then, the four panelists participated in an open discussion with the audience.

What Can the Health Sector Do to Help Local Efforts?

Rudolph remarked on the urgency of taking transformative action quickly, given that, she said, "We are blowing through what scientists call 'the carbon budget,'" and the reality is that each day action is not taken, the risk for more catastrophic climate change and its impact on future generations increases. Given this urgency, combined with the highly conservative political environment of Kentucky, Rudolph asked about the role of the health sector to help "move things along faster and scale things up."

Koetter responded that while the team at the University of Louisville has been a great help to the Office of Sustainability with respect to helping city officials understand the health data that have been collected, she encouraged a greater understanding of the effects of climate change on their part and among all health care professionals.

Hatic encouraged a more holistic perspective of health. Spiritual health is just as important as physical health, environmental health, and economic health, in her opinion. She said, "We can't dissociate one from the other."

Abbott offered that her message would be, "Be bold and really stand with communities that are facing these issues on the ground." She mentioned having attended a conference in Louisville on improving air quality in the city a few years before this workshop. The conference was attended by about 300 people from many sectors, and the day was packed with workshops, speakers, and good data and information about the health impacts of poor air quality. But the word "coal" was never mentioned, she recalled. She cautioned that if the only voices willing to talk about the evidence regarding what is actually causing some of the most significant health impacts are "voices on the margins of political power," it will remain very difficult to build political will. She encouraged people in positions of power, whatever power that may be, to "be bold enough together to say what is truly going on and what is needed."

Building Political Will

When asked by Catherine Baase of the Michigan Health Improvement Alliance what the state government's reaction to Empower Kentucky's People's Plan has been and what the next steps for the plan will be, Abbott replied that a primary goal will be to build political will. When the final plan is released, KFTC will be holding public briefings and presentations, as well as meetings with individual leaders. In addition to encouraging what she envisioned as incremental policy changes, her other hope for the plan is that it will be used by people who are thinking about running for office in 2018 or 2020. Candidates can "grab onto" some of the ideas and outcomes in the plan, she envisioned, and communicate to Kentuckians specific steps that can be taken to create jobs, save money, and improve health.

While on the topic of political will, Rudolph asked Koetter where the political will in Louisville is coming from for the Office of Sustainability work that Koetter described. Koetter credited the city's strong local leadership and awareness, including the mayor, who formed the Office of Sustainability when he entered office in 2011.

Mary Pittman of the Public Health Institute then asked about strategies for state-wide scaling up of some of what has been done in an urban environment given that, she said, "the culture of coal country is a bit different than an urban center." She agreed that part of the challenge is political will, but opined that another part is this culture shift. Hatic responded by telling a bit of her own personal history, including her move from Boston to Kentucky and her initial work with energy management programs in 30 school districts across eastern Kentucky. One of the things she quickly learned, she said, was that even if the adults sitting around the table did not buy into the "green movement," their kids did. She expressed a great deal of hope for the next generation, even in Appalachia, where people are beginning to recognize "the importance of the earth . . . and the fact that what I do to my planet, I am doing to myself."

"Culture eats strategy for breakfast every time," Abbott quoted, and agreed that culture matters and stressed the importance of messaging and organizing messengers. "Who is helping to bring the messages forward really, really matters," she said.

Engaging the Private Sector

When asked by John Bolduc, an audience member and environmental planner for the city of Cambridge, Massachusetts, how the business community is responding to work described by the panelists, Koetter described the response as "mixed." She mentioned that the several Fortune 500 companies that are headquartered in Louisville maintain corporate responsibility, but many local businesses have not been as receptive because of concerns about regulation. That said, however, some have been a little more receptive to the tree canopy work, for example, asking for help on where to plant trees on their property.

Hatic added that both she and Koetter sit on a local nonprofit called the Louisville Sustainability Council. Many of the larger businesses in Louisville are represented, so "they're showing up at the table," she said.

Natasha DeJarnett, an audience member from APHA, asked the panelists for insights into how to facilitate public–private partnerships for moving climate action forward. Hatic reiterated a key message from her presentation: "Check your agenda at the door." In addition to establishing good buy-in, she said, "it's all about relationship." She clarified that she meant not just relationships with others, but also relationships with ourselves and with the earth. She encouraged talking about "how [we] can work together through our shared values and our shared needs to create something that will be positive and sustaining."

Engaging Religious Networks

Gary Gunderson of the Wake Forest Baptist Medical Center and Stakeholder Health identified himself as a Baptist minister and mentioned that he works with many faith-based health care systems. He asked Hatic and Holmes about the capacity of the Landscape Audit for Sacred Spaces program to engage other religions or religious organizations, such as the Southern Baptist Theological Seminary, and particularly some of the more conservative religious networks.

Holmes replied that such engagement has to do with how issues are framed. The challenge, she said, is to find out what matters to the communities that you are working with and then frame the issue in those terms. For example, if the issue is water quality, while water quality has both upstream and downstream issues that need to be addressed, she asked, "What do people care about?" They care about the water in their glass, she answered. They are not necessarily thinking about the forested areas that are protecting the watershed that produces that water. In her opinion, it is not that other religious networks are not engaged. Rather, the challenge is to find out how they are engaged and how that engagement aligns with mutual goals. She added that although Pope Francis is considered by many to be the "poster child" for modern environmental consciousness of faith-based institutions, in fact, it was the evangelicals who "put faith-based environmentalism on the map." Their environmentalism can be traced back to St. Francis and Thomas Aquinas, perhaps even Jesus, Holmes suggested. Thus, she emphasized, a foundation exists for these partnerships to be formed.

Hatic described the environment as a "unifying" topic, not an issue that separates, and that conversations leading up to the Landscape Audit for Sacred Spaces were not new. She remarked that much of the dialogue at the Festival of Faiths event in Louisville, which the Center for Interfaith Relations has been hosting for 22 years, revolves around the environment. She added that the Center is constantly reaching out and inviting everyone in the community to the table and that responses and audience members are different from year to year. In fact, she said, the Southern Baptist Theological Seminary was represented at a talk hosted by the Center a couple of years ago. "The seeds are being planted," she said, but "it takes a little while for them to germinate."

A Just Transition from Traditional Energy

Finally, much of the discussion at the end of this first panel revolved around the impacts of transitioning from traditional energy and the concept of a just transition. First, a member of the workshop audience, Kelly Dennings, asked the panelists how their organizations were addressing the transition of Kentuckians from the traditional energy sector with respect to mental health and other components of health.

Koetter replied that the Louisville Office of Sustainability's goal is to reduce energy use per capita by 25 percent by 2025. As she had said during her talk, she mentioned again that hopefully, they will be engaging a biostatistician to help them understand how the improvements in air quality resulting from this reduced energy use will impact health. She encouraged putting health first when promoting energy goals as a way to defray some of the political sensitivity around the issue.

Since 2009, Kentucky has lost about 13,000 coal mining jobs, Abbott responded. The mining communities were already among some of the nation's poorest, even when the coal industry was doing well, she said, so the loss of the sector and the loss of the best paying jobs has hit the communities hard. "They are struggling on multiple levels," she said. The hit includes not just loss of revenue, but population loss and mental health impacts, including drug abuse issues. Their resilience to deal with some of these impacts is greatly challenged. She noted that the former federal administration made some small, but important, investments, and suggested that much more could be done by both the federal government and the state of Kentucky to provide comprehensive support for these communities. For example, KFTC is working to pass federal policy that would direct about $1 billion to land and stream restorations, which would have ecological benefits and would be a job-creating force. In addition, Abbott underscored the need for support not just for the coal workers, but also for female heads of households who have never had good economic opportunities in these communities. She said, "Thinking about a just community has to include the directly affected workers, but also the broader community in which they live."

Dick Zimmer of Zimmer Strategies asked Abbott to elaborate on what she meant by "just transition" and what other opportunities exist for an unemployed coal miner besides stream restoration. Abbott explained that the term "just transition" comes from union movement language, specifically discussions about plant closures. In the context of coal mining regions in central Appalachia, KFTC is working to broaden the term to include the whole community in addition to the affected workers. She reiterated that the economies in these communities experience some of the most persistent and deepest poverty in the nation. She said, "The whole region is in need of a just transition."

KFTC has developed a number of principles and a set of policies that illustrate what a just transition would look like, Abbott continued. "We don't have all the answers," she said, but there are some things that could be done today to support a just transition. These include supporting some of the foundational parts of any healthy community, such as a strong health safety net, strong local schools, opportunities for meaningful job training, and investment in job creation. Importantly, she said, a just transition also includes providing job-sustaining support for workers who, for example, have been working on the railroads for 17 years, but have recently been laid off because there is so much less coal being transported, and therefore will not receive their 25-year guaranteed pension. There has to be a public investment in supporting these workers to bridge them to the point of retirement, she argued.

Although the economy as a whole is not healthy in eastern Kentucky, there are opportunities, Abbott said, and many unemployed coal workers have skills that can be redirected toward other careers that can provide family-sustaining wages. Land and stream restoration is one opportunity, which has as its goal repairing some of the legacy of extraction. Other opportunities exist in sustainable forestry, renewable energy growth, health care, tourism, and arts and crafts.

Patricia Martz, a member of the workshop's remote audience from Edmonton, Alberta, offered two examples of a just transition outside of Kentucky. The first was tobacco farmers in the southern United Sates who no longer grow tobacco, rather chickpeas, and who also, as a "bonus," the audience member noted, no longer chew tobacco, rather chickpeas. The second example she cited was the closure of the logging and timber industry in Leavenworth, Washington, where, in 1962, Project LIFE (Leavenworth

Improvement for Everyone) was formed in partnership with the University of Washington to investigate strategies to revitalize the struggling logging town and where the old logging and timber jobs and economy were diverted into tourism.

Zimmer also asked about the role of nuclear power in the Empower Kentucky plan. Abbott explained that Kentucky has banned nuclear power since the 1970s, and while there is an effort in the legislature to remove the ban, Empower Kentucky did not consider it. It was mentioned occasionally, Abbott said, in KFTC's conversations around the state, but "let's never advance it" is mentioned as frequently as "let's advance it." She mentioned that one of the state utilities, Kentucky Power, recently analyzed its next 15 years and ruled out nuclear power as an option because of its expense.

4

Regional Perspectives from the Midwest

Panel 2 moderator Surili Patel offered opening remarks on the reality that not everyone is equally resilient to the health impacts of climate change. She encouraged workshop participants to keep this in mind as they discuss policies and strategies for responding to these impacts. Particularly vulnerable populations include children, the elderly, tribal communities, low-income communities, and communities of color. She reiterated a comment that keynote speaker Jonathan Patz made earlier about the importance of listening and encouraged "going into those communities" to gain an understanding of their cultures and histories.

The focus of this second panel was on strategies for responding to climate change that are being implemented by health agencies and organizations in the Midwest. First, Paul Biedrzycki from the City of Milwaukee Health Department emphasized the important role of local public health officials in responding to climate change and discussed how this role is playing out in the City of Milwaukee. Then, Jeff Thompson from Gundersen Health System in Wisconsin, described Gundersen's goal to become 100 percent energy independent through conservation and the use of renewable energy. This chapter summarizes these two presentations. Key points made by each panelist are listed in Box 4-1.

In the open discussion with the audience at the end of the panel, topics addressed included framing climate change as a health challenge; climate change awareness among health professionals at the national and international levels; what is unique about Gundersen that has allowed it to "push the envelope" with its energy system work; and how health agencies and organizations working on climate change strategies can build trust with

---

BOX 4-1
Key Points Made by Individual Speakers

- Local public health departments are at the front line of climate change preparedness and response, yet, according to a 2014 National Association of County & City Health Officials survey, less than 5 percent of local health departments have adequate capacity or capability to plan for the impacts of climate change. This inadequacy pushes these departments to seek innovative collaborations within their communities. (Biedrzycki)
- Examples of successful local climate change adaptation strategies being implemented in Milwaukee include a community garden rainfall harvesting program; changes in both the reporting of Lyme disease and risk messaging about Lyme disease to health care providers and to the public; engagement with large corporate entities to ensure business continuity in the face of different climate change scenarios; and development of a heat vulnerability index. (Biedrzycki)
- To "truly live" its mission to keep people healthy, in 2008, Gundersen Health System embarked on an ambitious energy system program to decrease pollution while simultaneously reducing operating cost. At the time, Gundersen was emitting 435,000 pounds of particulate matter, more than 70 million pounds of CO_2, and nearly 2.5 pounds of mercury annually. (Thompson)
- Starting with energy conservation, followed by wind energy generation, a landfill bio-gas project, and several other renewable energy efforts, Gundersen's fossil fuel emissions have been reduced by more than 90 percent. In 2016, only 11,000 pounds of particulate matter, 1.6 million pounds of CO_2, and 0.16 pounds of mercury were emitted. Not only have fuel emissions improved, but Gundersen's energy expenses are below the 2008 spending level. (Thompson)

---

communities. The discussions on each of these issues are summarized at the end of this chapter.

LOCAL PUBLIC HEALTH PERSPECTIVES: CLIMATE HEALTH PREPAREDNESS AND RESPONSE[1]

Paul Biedrzycki provided a local health department's perspective on climate health, including results from a 2014 National Association of County & City Health Officials (NACCHO) survey of local health officers ("Are We Ready? Report 2"), and four examples of climate health adaptation strategies from the City of Milwaukee. The survey results, he said,

---

[1] This section summarizes information presented by Paul Biedrzycki, director of disease control and environmental health, City of Milwaukee Health Department, Wisconsin.

characterize the capacity and capability of challenges faced by local health departments across the country.

Milwaukee

The City of Milwaukee has about 600,000 people. Its larger metropolitan area, which now encompasses five counties, totals about 2 million people. Biedrzycki described Milwaukee as a majority/minority and highly segregated community, with about a 40 percent African-American population centered mostly on the north side of the city. There is also a growing Latino population on the south side. Both of these populations, he said, are among the lower socioeconomic status populations in the city.

Being on Lake Michigan, Milwaukee markets itself as the "freshwater capital." Sometimes the mayor also refers to it as the "freshwater city of the Great Lakes" or the "fresh coast city of the Great Lakes," according to Biedrzycki. The city's freshwater center occupies a growing presence in the city's research, technology, and business.

The Role of Local Public Health

Biedrzycki listed several ways that local public health agencies are well positioned to deal with climate health issues and emphasized that it remains well positioned, even as its capacity and capability are extremely limited these days. First, community health assessment, assurance, and policy development are part of the mission of local public health. Local public health agencies also maintain a focus on health disparities. Biedrzycki referred to Patel's opening remarks on health equity and added that climate change exacerbates existing health disparities and inequities.

Another factor contributing to the well-positioned role of local public health, in his opinion, is that public health agencies speak articulately on the economic, social, and political intersections related to public health, hence the public health-in-all-policies approach that is now being advocated at many levels.

Finally, Biedrzycki pointed to local public health's prevention-oriented approach to health care, which, he said, is the most cost-effective approach and "fundamentally reasonable."

Community Social Capital

In addition to the many other physical and mental health impacts of climate change that were previously identified by other workshop speakers, Biedrzycki underscored the importance of what he referred to as "community social capital." Interpersonal aggression, violence and crime, and social instability are often overlooked or underemphasized, he noted, with discussions on climate health issues tending to focus instead on some of the better known impacts on food, water, and air.

Are We Ready?

The first NACCHO "Are We Ready?" survey was conducted in 2008 (NACCHO, 2008), and the second in 2014 (NACCHO, 2014). The surveys measured perceptions of local health officers and departments on climate health, including its underlying causes, human health impacts, and the ability of local health officers and departments to respond to these impacts. The majority of NACCHO's 2,300 members represent small to medium-sized health departments, Biedrzycki noted. Not surprisingly, in his opinion, the 2014 survey showed that nearly 8 out of every 10 local health directors recognized that climate change is occurring, regardless of underlying attribution, compared to 6 out of every 10 Americans.

Among other health-related scenarios that survey respondents were quizzed on, Biedrzycki pointed to one in particular that stood out to him: disruption of health services for people with chronic conditions. That people who are chronically ill are often impacted when power or other key infrastructure components are disrupted is often overlooked, Biedrzycki noted.

The take away from the survey, in Biedrzycki's opinion, was that fewer than 5 percent of local health departments had programs to educate the public about climate change and potential impacts. Similarly, only 19 percent have ample expertise and 8.4 percent agree that they have sufficient resources to protect residents from health impacts associated with climate change. This lack of capacity or capability, Biedrzycki said, "pushes us to more innovative, cost-effective collaborations within the community."

Local Climate Health Adaptation Strategies: Examples from Milwaukee

Biedrzycki discussed in detail four climate health adaptation strategies implemented by the City of Milwaukee that he thought had been particularly effective. The first was a community garden rainfall harvesting program funded by a small grant from the Public Health Institute. One of the program's two community gardens serves the largest male homeless population in Milwaukee, known as the "Guest House," while the other, "Alice's Garden," provides produce to local markets and restaurants. Both gardens are located in the low socioeconomic area of the north side of Milwaukee. The many co-benefits of rainfall harvesting at these gardens include not just food security, but also reduced runoff pollution, water conservation, and increased visibility to climate change issues. Having climate conservation within the broader community and in these types of environments in particular is "incredibly important," Biedrzycki said. Overall, he described the program as win–win, low cost, not a heavy lift, creative and innovative, and a way to bring visibility to the issue.

A second example involves Lyme disease reporting and response within the City of Milwaukee. Biedrzycki described Lyme disease as a very serious disease resulting in joint pain and a rash. It can be treated with antibiotics, but may cause significant cardiovascular and neurological effects later in life. Centers for Disease Control and Prevention (CDC) data show a marked increase over the past 25 years in Lyme disease reporting nationwide, particularly in the northeast and mid-Atlantic states. Biedrzycki stated that many experts believe that the multifold increase in Lyme disease incidence in both the Midwest and northeast is connected with climate change. The disease is transmitted by the black-legged tick, or deer tick, with white-footed mice serving as the reservoir and deer the transporter. As these animals' ecosystems change, so does their geographic distribution, he explained. In Wisconsin, these ticks, once seen only in the northwest quadrant, are now distributed throughout the state. This geographic redistribution has required a change in both the reporting of Lyme disease and the risk messaging to health care providers and the public on prevention and treatment. For Biedrzycki, this change demonstrates "an adaptive technique associated with potential transmission of communicable disease in a jurisdiction."

A third example of an innovative approach to climate change adaptation is the City of Milwaukee's engagement with what Biedrzycki described as "corporate America in Milwaukee." Milwaukee is the home of several large corporate entities, including MillerCoors, Harley Davidson, and Northwest Mutual Life Insurance Company. The city has been able to engage these companies in public health emergency planning in response to several threats, particularly the post-September 11 bioterrorism threat and the threat of pandemic influenza.

The focus of the city's current engagement regarding climate health is on business continuity in the face of different climate change scenarios, such as a power outage during extreme heat or a prolonged heat event, a drinking water shortage, or a disease outbreak. Biedrzycki recalled that Milwaukee was the epicenter for the 1993 *Cryptosporidium* outbreak, which he noted was the largest outbreak ever recorded in North America. During that outbreak, boiled water advisories were in effect for multiple weeks, bringing to light the fact that while water appears to be plentiful in Milwaukee, there are climate-related scenarios that can impact the potable water supply. As another example, he mentioned a more recent toxic algae outbreak in Toledo. He expressed uncertainty regarding whether much of corporate America nationwide is thinking about this dimension of climate change, even though these outbreaks affect the entire community, including the workforce. Bringing the Milwaukee businesses up to speed on these issues, particularly the risk of communicable disease in the face of a climate event, via a focus group, presentations, and surveys has, he said, "really attracted their attention."

In addition to businesses' concerns about continuity of operations, Biedrzycki pointed to their strong interest in green marketing as well. He observed that the emerging millennial workforce has strong sentiments about working for corporations that are sensitive to sustainability and climate change in general.

The fourth and final example that Biedrzycki used to illustrate effective strategies being used in Milwaukee was the city's development of a heat vulnerability index. When the index was developed in 2012–2013, Biedrzycki considered it a "groundbreaking" effort, given that Milwaukee was one of only five or six cities with one at the time. He explained that a heat vulnerability index overlays natural and built environmental, demographic, socioeconomic, and health behavior variables predictive of heat susceptibility. Milwaukee's index, which has about 20 such variables, shows that the greatest risk for negative health impacts due to extreme heat is in the poor areas of the city. Biedrzycki remarked that the development of the index, which was funded by a Building Resiliency Against Climate Effects (BRACE) grant, allowed the city health department to "jumpstart" strategic decision making around heat health, for example, where to place cooling centers, how long to keep public swimming pools open, and where to target messaging to vulnerable populations.

Equitable Adaptation and Community Engagement

Biedrzycki concluded with some remarks on the importance of equitable adaptation and community engagement. First, he referred to a 2016 summit held at the Georgetown Climate Center in Washington, DC, on opportunities for equitable adaptation in cities. At the summit, some key tenets associated with equitable adaptation were identified. He stressed the importance of these tenets given what is known about how even small impacts of climate health exacerbate existing inequities. The tenets include, for example, building trust between government and the public and leveraging community-driven networks and resources to address climate health scenarios.

Finally, he mentioned a community engagement toolkit for Wisconsin, funded by the BRACE grant, that is now being integrated into local health department emergency planning, especially with respect to climate change scenarios (Wisconsin Climate and Health Program, 2016). Additionally,

NACCHO's 12 steps to operationalize climate change,[2] which were meant to serve as guidance for local health departments, underscore community engagement for the development, implementation, and evaluation of climate adaptation strategies.

Moving Forward

In summary, Biedrzycki listed the steps that have been built into Milwaukee's climate health planning and response:

- Build community awareness and inspire action, especially among local policy makers, such as the mayor's office (i.e., making the economic argument compelling and convincing to policy makers);
- Leverage community networks and resources, particularly given that "the money just isn't there right now," according to Biedrzycki, and be creative about programs in environmental health, communicable disease, and emergency preparedness that are already on the ground;
- Incorporate climate adaptation into the economic sustainability and urban planning framework; and
- Advocate for climate equity.

PROTECTING THE HEALTH AND WELL-BEING OF COMMUNITIES IN A CHANGING WORLD[3]

Gunderson Health System is an integrated health care system with a number of hospitals and clinics across Iowa, Minnesota, and Wisconsin. Jeff Thompson provided an overview of Gundersen's energy system work, including how and why they took on the climate change and sustainability issues that they did. One of the great leverages that Thompson and his colleagues used to initiate Gundersen's energy system work and communicate not only with the board, but also community partners, was the purpose of the organization: "Our purpose is to bring health and well-being to our patients and communities." They described "health and well-being" broadly to also include financial health, social health, and environmental health.

---

[2] The 12 steps are described at http://www.naccho.org/uploads/downloadable-resources/NA634PDF-12Steps.pdf (accessed May 9, 2017).

[3] This section summarizes information presented by Jeff Thompson, executive advisor and chief executive officer emeritus, Gundersen Health System, La Crosse, Wisconsin.

Arguing for Energy Independence (Without Mentioning Climate Change)

Regarding why a health care organization would want to become involved in activities related to climate change and sustainability, the answer, Thompson said, is "the incongruity." He explained that the mission of Gundersen is "We will distinguish ourselves through excellence in patient care, education, research, and improved health in the communities we serve." Yet, he said, "we're supposed to be keeping people healthy, but we're killing them with our pollutants."

According to the U.S. Department of Energy, hospitals are 2.5 times more energy intensive than schools and other commercial buildings. Given that the health care sector accounts for approximately a sixth of the economy, its energy impact is huge. Pharmaceuticals and hospital equipment have an even bigger carbon footprint, Thompson noted. To add to this, in 2008 when Thompson and his Gundersen colleagues began their efforts to change Gundersen's energy use, they believed that energy prices would continue to rise. Although what Thompson described as the natural gas "glut" has flattened a bit, today they continue to believe that prices will eventually rise again. Finally, Thompson and partners also believed that reducing waste would result in greater savings.

Thompson clarified that he and his colleagues did not set out to "sell climate change" to the board. Even mentioning climate change, he said, especially back then, would have "been a fail." He admitted, however, that if they were to have asked him about climate change, he would have told them that there is better evidence for climate change than for the majority of treatments Gundersen provides in its institutions. Nor was climate change the discussion back around 2008, Thompson explained. Rather, the discussion revolved around how to decrease pollution that is causing harm to people, how to do so while decreasing operating cost, and how to improve the local economy and stop sending money out of the state (e.g., importing coal from Wyoming for electricity and natural gas from Texas for heat). It was also believed that changing the energy system of Gundersen would help the organization to "truly live" its mission, as opposed to "just giving lip service."

Finally, some of the inspiration for their efforts came from *Natural Capitalism*, which Thompson described as a book about ignoring conventional wisdom that one must choose between jobs or the environment (Hawken et al., 1999). In making their case to the board, Thompson and partners argued that their proposed energy plan would be the best use of Gundersen's savings; that Gundersen would receive a good return on investment; it would be safer than the stock market (i.e., again, Thompson noted, this argument was put forward in 2008); that the local investment required of the plan would make for great public relations; that there were available partners in the community (e.g., Thompson commented on the faith community in particular); and that Gundersen's early conservation projects had been successful. Regarding the last point, Thompson noted that, by spending $2 million in the first 2 years of the program, Gundersen realized savings of $1.2 million in every following year.

Another argument that can be made when trying to build coalitions of people, Thompson said, is "to face the brutal facts." In 2008, Gundersen was emitting 435,000 pounds of particulate matter, more than 70 million pounds of CO_2, and nearly 2.5 pounds of mercury every year. Regarding the mercury emissions, Thompson explained that one gram of mercury pollutes hundreds and hundreds of acres of water, so pounds pollute tens of thousands of acres. He noted that the reason pregnant women and children are advised not to eat fish from lakes in upstate New York is that the water is contaminated with coal debris from Wisconsin and elsewhere. "So this is truth," he said, referring to these emission figures from 2008. "This is what we're doing."

Achieving Energy Independence: Gundersen's Strategy

Gundersen set a goal in 2008 to be 100 percent energy independent by 2014, with a key component and the first step being to decrease energy use. Thompson emphasized that energy conservation does not require Congress passing a law or the generation of new energy. It means simply saying, "We are going to start conserving." As a result of its initial conservation efforts, Gundersen Health expected to achieve in 2017 a 59 percent improvement in energy efficiency compared with 2008 (see Figure 4-1).

After its initial conservation phases, most of the next steps in Gundersen's effort to gain energy independence have involved the generation and use of renewable energy, beginning with wind energy generation and a landfill bio-gas project (see Figure 4-2). Regarding the latter, the county installed a pipe from the county landfill to Gundersen's northern clinic so that Gundersen could buy and use the gas from the landfill. Today, Gundersen uses the gas to heat, cool, and power an entire campus, while the county receives $250,000 per year. Prior to this landfill bio-gas pipe being built (i.e., by Gundersen), that gas, Thompson said, was "just being flared into the atmosphere." The project, Thompson said, has been good for Gundersen by lowering the cost of care; good for the community by providing a source of income so that the county does not have to raise taxes; and good for the environment because it displaces other energies, namely coal, which is a major source of electricity in the upper Midwest.

As illustrated in Figure 4-2, in addition to conservation, wind energy, and the landfill bio-gas project, Gundersen has also invested in geothermal fuel, a high-tech biomass boiler, a dairy bio-gas project, and a variety of

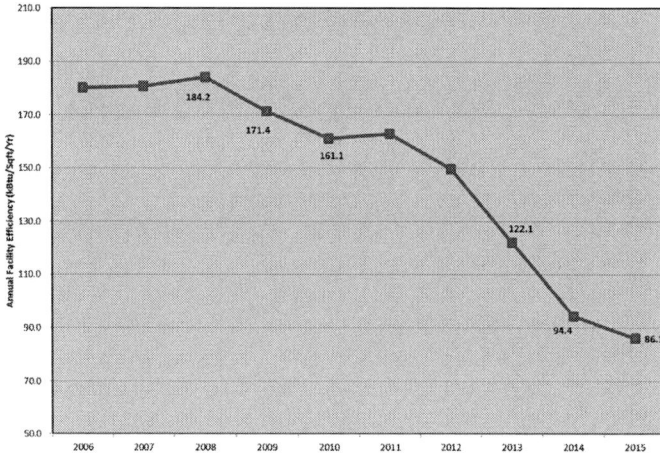

FIGURE 4-1 Gundersen Health System (GHS) energy efficiency: change in GHS energy efficiency since 2008, as measured by the difference between aggregated facilities' energy use intensity and utility-purchased energy.
SOURCE: Thompson presentation, March 13, 2017; reprinted with permission from GHS, 2017a.

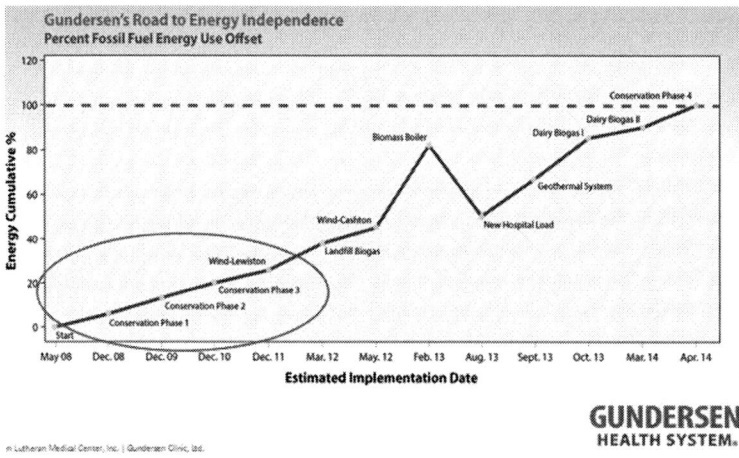

FIGURE 4-2 First U.S. health system heated, powered, and cooled by local renewable energy.
NOTE: Strategies used by Gundersen, since 2008, to become energy independent.
SOURCE: Thompson presentation, March 13, 2017; reprinted with permission from GHS, 2017b.

other projects. Thompson noted that the financial returns for each of these projects vary, but emphasized that all of these forms are renewable energy. For example, a biomass boiler uses hardwood chips that otherwise would just rot into the ground, while dairy bio-gas not only generates electricity, but also prevents both methane from entering the atmosphere and phosphorous from entering lakes and triggering massive algae blooms.

Not all of Gundersen's renewable energy efforts have been successful, Thompson acknowledged. Their brewery bio-gas attempt, for example, failed. The goal of that project was to generate electricity from beer gas. But the company they had partnered with switched to working on other projects. Plus, the company used a method that relied on sulfuric acid, which damaged the bio-gas engine.

Reduced Fossil Fuel Emissions

Again, the goal set back in 2008 was to become 100 percent energy independent, Thompson continued. He remarked that while 2016 saw only 90 days of energy independence, Gundersen's fossil fuel emissions have been reduced by more than 90 percent. Specifically, in contrast to the 72 million pounds of CO_2 emitted in 2008, only 1.6 million were emitted in 2016, representing a 98 percent reduction; in contrast to the more than 435,000 pounds of particulate matter emitted in 2008, only 11,000 were emitted in 2016, representing a 97 percent reduction; and in contrast to the more than 2 pounds of mercury emitted in 2008, only 0.16 pounds were emitted in 2016, representing a 94 percent reduction.

Increased Return on Investment

Not only have fossil fuel emissions improved, but Gundersen has also made money on these efforts by changing its investment portfolio. Specifically, Thompson explained, by removing 5 percent of savings (i.e., from cash, treasury bills, bonds, and stocks) and reinvesting these funds into Gundersen's energy infrastructure, their return has increased from 5–6 percent to 10–12 percent. Reinvesting what he described as "stranded assets that are sitting there that could be used for other things" is, he said, "a different level of thinking."

As a result of these efforts, Gundersen's energy expense remains below the 2008 spending level (i.e., $5,016,000 in 2015, compared to $5,348,264 in 2008) and well below what would have been spent if they had maintained "business as usual" (i.e., projected $7,688,000 in 2015).

The local economy has been boosted as well, Thompson noted. "So you can't tell me it's jobs versus the economy," he said. He added that beyond improved health and savings for the health system, staff pride has blossomed, Thompson continued. The environment and sustainability are part of the value set of college graduates today, and thus Gundersen has been attracting employees because of its energy program.

Concluding Thoughts

Thompson concluded by reiterating that Gundersen did not set out to "save the climate." Rather, its pitch revolved around health and the economy. He encouraged workshop participants to visit Gundersen's website (www.gundersenhealth.org) to learn more about the mechanics of what it has done and to read his book, *Lead True*, on value-based decision making (Thompson, 2017).

DISCUSSION

Following Thompson's presentation, panel moderator Patel commented on the emphasis that both Thompson and Biedrzycki placed on partnerships. "Understanding climate change is such a big feat," she said, "that we can't do it on our own." She referred to Biedrzycki's statement that less than 5 percent of local health departments, who are on the front line, have the capacity to respond to climate change. That, she said, is why partnerships are so important. In addition, she appreciated Thompson's emphasis on the reality that calling the challenge "climate change" does not matter. That is, regardless of the language used, she said, "the solutions are where we are going to benefit the most." The two panelists then addressed several questions from the workshop audience.

Framing Climate Change as a Health Issue

Linda Rudolph observed that both Jonathan Patz and Georges Benjamin had emphasized the importance of framing climate change as a health issue earlier in the morning, yet Thompson, in this panel, spoke about *not* leading with climate change. Additionally, she remarked that based on having perused many local health department websites, including Milwaukee's, climate change is not prominent in any but a small number of local health department communications. She asked the panelists how they reconcile these different perspectives and what they think the responsibility of health leaders is with regard to raising the level of awareness among both policy makers and the public that climate change is one of the greatest health challenges today.

Thompson replied that in 2007–2008, he realized that what he wanted to do with Gundersen's energy system would require millions of dollars of investment. If he had told the board he needed millions of dollars to mitigate climate change, he said, "it would have been a non-starter . . . they would have just shut me down." He had to identify where their values overlapped with his. Again, he said that if they had asked him about climate change, he would have been very clear about climate change and its devastating health effects. In fact, he has spoken out about climate change at a number of places in his community, as well as regionally and around the country. He mentioned, again, that there is better scientific evidence for climate change than for most medical treatments in the United States. But to move forward within his organization, he had to frame the problem in a different way.

Biedrzycki replied that while he had mentioned during his presentation that local public health departments were on the front line, it is actually the private health care providers who are on the front line. Private health care, in his opinion, has an essential role in broaching the climate conversation within the community. The reason climate change is not mentioned on Milwaukee's health department website, or on any of the Wisconsin municipal websites, is because it is not politically popular at this time to even say "climate change" given the state's current governor and some of the state's municipal leaders, he said. He called for a way to navigate conversations with policy makers and other key community leaders and champions so that the impacts of climate change can be addressed in a manner that creates a win–win solution—in other words, one that results in cost-effective economic solutions that are good for both the community and business. "We recognize the challenge," he said, but "have yet to come up with an easy solution."

Climate Change Awareness Among Health Professionals at the National and International Levels

Jonathan Patz asked Biedrzycki about the pervasiveness of climate change awareness among members of NACCHO. He asked if climate change is "on the table" nationally. Additionally, he asked Thompson whether, based on Thompson's recent international trips and audiences, climate change is gaining traction among health ministries.

Climate change is gaining traction at NACCHO, Biedrzycki replied, particularly among the emerging, younger workforce. For example, NACCHO now has a climate health workgroup. But most of the discussion has revolved around what is being done locally. He acknowledged that local case studies can be stimulating, particularly with respect to their demonstration of unique partnerships and collaborations with nontraditional stakeholders and of ways to shift the dialogue outside of what Biedrzycki called the "groupthink" around climate health issues. However, he called for a comprehensive strategic plan and for strong national leadership as well. Currently, there is no cohesive, comprehensive strategic plan that links what local public health agencies are doing with what is being done by state and federal agencies. Noting that the intention was not to be overly critical of CDC and that he was aware that the American Public Health Association was a strong advocate and promoter, in his opinion, there is no strong national leader or champion. Until a national leader emerges, he said, "I think we are going to struggle with developing this comprehensive plan that interlinks us horizontally and vertically across our communities nationwide."

Thompson added that the opportunity is not dissimilar to other times in history when governments "dragged their feet" for political, financial, or other reasons. He said there are many examples of middle-sized organizations, like Gundersen Health System, moving forward on their own to make a change. Additionally, he mentioned a group of mayors, some states, and a medical society group that are all moving forward despite the federal government's stance. He also noted that internationally, even in China, most of the world recognizes the urgency of the challenge. He said, "I think the movement forward is to say, 'The cause is so important that we will build coalitions of people with similar values . . . and a similar goal.'"

What Is Unique About Gundersen Health System?

David Kindig asked Thompson what the state of acceptance is regarding Gundersen's energy use and investment program and what is unique about Gundersen that has allowed it to "push the envelope" as it has. Thompson replied that the state of acceptance is "on a steady upward curve." A few years ago, Thompson and colleagues developed the Healthy Hospital Initiative. In the course of just 3 years, membership grew from about a dozen organizations to about 1,400 hospitals and health systems across the country reporting on how they are doing with energy conservation, waste management, and water. There is escalating interest, he said, both with respect to the effects on health and the fact that this can be done in an economically viable way. However, he did not expect anyone to adopt the Gundersen model exactly, because different places have different environments and costs. For example, places that have more sun and lower photovoltaic costs would likely benefit more from solar energy than from diary digesters. He added that, in his opinion, while acceptance is happening, it is not happening quickly enough. He suspected that it will take a generational change in leadership to make doing what Gundersen has been doing "more natural."

Regarding the uniqueness of Gundersen, Thompson remarked that one of the things that is appreciated at Gundersen is long-term thinking. There are no political cycles at Gundersen, he said, nor do they provide yearly bonuses for executives. Thus, their tolerance for risk is a little higher than it might be otherwise. But most importantly, in his opinion, is that what they have done is consistent with a values-based approach, one that would not shift with every quarterly report. Initially, finance was treated as a tool, not a goal. This does not mean that they did not have to hit financial targets. They did, Thompson said. But treating those targets as a tool to accomplish a mission, rather than as a goal for either the organization or individuals within the organization, made it easier initially to engage government, nongovernmental organizations, business, and other partners.

Biedrzycki added that, in his opinion, what is missing in the current narrative is what success looks like from a climate-healthy business economy and how that success intersects with national security and economic goals. Until that narrative is written and that intersection is defined, which he said is something typically done by the citizenry, it will be difficult to establish a good strategic plan and predict a future trajectory.

Building Trust

Patel asked Biedrzycki to provide an example of the trust building that the City of Milwaukee has done with communities. Biedrzycki defined trust building as genuine and authentic relationship building. In his opinion, it is at the core of many public health programs. But it is difficult to achieve, he said. It takes time, face-to-face encounters, and development of win–win situations that are mutually consensual and agreeable. He referred to Milwaukee's rainfall harvesting project, which he had described during his presentation, as an example of a win–win relationship on many fronts. For example, the garden that is located in a male homeless shelter represents a win for public health by engaging that part of the community and a win for the community by providing food security for what is a very vulnerable population. Additionally, Biedrzycki pointed out, the project has supported Milwaukee's award-winning HOME GR/OWN initiative by repurposing vacant properties into community gardens. The project's success required agreeing, beforehand, on what constitutes successful outcomes for each partner. Building the trust to do that requires time. "It's not prescribed," he said, "It's developed."

Sanne Magnan observed that the rainfall harvesting project appears to have built what Biedrzycki had referred to in his talk as "community social capital." She asked whether it is a lack of such capital (i.e., decreased community cohesion, increased social instability) that affects the ability to work on climate change or if it is working on climate change that impacts this capital. Biedrzycki responded, "Both." He reminded the workshop audience that Milwaukee is a highly segregated city with great racial divides. As they have done in many urban communities, these divides have created tensions among the government—in Milwaukee's case, the police department—and its citizenry. This tension, he said, can work against building trust. On the other hand, when meaningful seminal partnerships are formed, as they were with the rainfall harvesting project, then these tensions can be assuaged and trust can be built.

Also related to the issue of building trust, Pamela Russo of the Robert Wood Johnson Foundation commented on the clear win–win situation demonstrated by Gundersen's investment in the community. She then asked first, whether any of the work that Thompson discussed was counted in the organization's community reporting, and second, whether there were any issues or hindrances that made it difficult for Gundersen to move forward in any particular localities or states. Thompson was unsure which of their many community partnerships were reported. Regarding the second question, he referred to the "morass" of rules and regulations, which are different in every state, and the complexity this creates. Despite this challenge, he said they had good luck working with the state governments, and with the federal government as well. He said, "We're almost always able to find people who are inspired enough to roll up their sleeves and say, 'Let's find a way,' rather than just say, 'No, you can't do that.'"

5

Regional Perspectives from the Northeast

In Panel 3, moderated by Paul A. Biedrzycki, the geographic focus shifted to the Northeast. First, Celia Quinn discussed emergency preparedness in New York City (NYC), with a focus on the city health department's experience with Hurricane Sandy and the expectation that extreme weather events will be more frequent in the future. Next, Matt Cahillane discussed New Hampshire's use of the Centers for Disease Control and Prevention's (CDC's) Building Resilience Against Climate Effects (BRACE) framework and described several local-level, climate-related public health projects under way across the state. Lastly, Kristin Baja provided an overview of Baltimore's unique approach to climate planning, one that incorporates climate change into the city's generalized all-hazard mitigation plan and that focuses on equity. This chapter summarizes these three presentations. Key points made by the panelists are presented in Box 5-1.

In the open panel discussion with the workshop audience, a range of issues were addressed: collaboration across the different sectors of the NYC health care system; mortality and other outcomes of the Hurricane Sandy evaluation; alert, disaster, and fatigue; an elaboration of Baltimore's incorporation of climate planning into its All-Hazards Mitigation Plan; funding for all of the various programs discussed by the panelists; and whether and how the different planning approaches discussed by the panelists had been scaled from approaches others had taken and whether and how the approaches could be scaled up even further, for example, to a national level. The discussions around each of these issues are summarized at the end of this chapter.

BOX 5-1
Key Points Made by Individual Speakers

- Health care system resiliency should be a primary goal of health care system emergency preparedness. Several key concepts highlighted by the experience with Hurricane Sandy (e.g., the complexity of evacuation decision making) are being translated into actions to develop a more resilient New York City (NYC) health care system (e.g., the NYC health department is collaborating with the local hospital association to address some of the challenges of a large-scale evacuation). (Quinn)
- During a disaster, the health care system breaks in places where it is already weak. Thus, health care system emergency preparedness must go beyond making plans or even training and exercising those plans. Ensuring the best possible outcomes requires the system to be moving toward a resiliency framework with the goal being to make the day-to-day system stronger and offering consistent, high-quality services to everyone in all communities. (Quinn)
- New Hampshire's state public health department uses the Centers for Disease Control and Prevention's Building Resilience Against Climate Effects (BRACE) framework for its climate and health projects. The framework is worth more than any final plan (i.e., it does not just "sit on a shelf"), and it promotes collaboration (e.g., between the state and university scientists). (Cahillane)
- As a result of work done during the "projecting local health burdens" phase of the BRACE cycle, New Hampshire has identified three priority climate-related health concerns: (1) heat stress, (2) habitat change (with an emphasis on its impact on ticks and Lyme disease), and (3) extreme weather-related injuries and resilience. (Cahillane)
- Baltimore's approach to climate planning is unique because it uses the federal government's required All Hazard Mitigation Plan, but incorporates climate change into this plan. Additionally, the city considers not just "shocks" (i.e., single-event disasters, like a flood or hurricane), but also "stressors" (i.e., factors that pressure Baltimore on a daily or recurring basis, like violence). (Baja)
- Baltimore's equity-based approach to climate planning involves acknowledging the city's past, namely its history of intentional racism and segregation; prioritizing neighborhoods with the greatest vulnerability and historic disinvestment; and actively listening and engaging with members of the community. (Baja)

HEALTH CARE EMERGENCY PREPAREDNESS
IN NEW YORK CITY[1]

Celia Quinn opened by stating that the bulk of the work that NYC's Office of Emergency and Preparedness Response does to support the city's

---

[1] This section summarizes information presented by Celia Quinn, director, Bureau of Health Care System Readiness, New York City Department of Health and Mental Hygiene Office of Emergency Preparedness and Response, New York.

health care system to respond safely and effectively in all emergencies is directly informed by the city's experience during Hurricane Sandy and the expectation that extreme weather events will be more frequent in the future.

She emphasized that she would be looking through the lens of health care system resiliency to tell the Hurricane Sandy story, including impacts of the storm, how the health care system responded both immediately and during ongoing recovery, and how the entire experience changed the thinking about health care system preparedness in NYC. She emphasized the importance of a resiliency approach in today's increasingly unpredictable climate.

Additionally, she noted that she would first be drawing on the NYC Special Initiative for Rebuilding and Resiliency report, *A Stronger, More Resilient New York*, which was part of the mayor's PlaNYC initiative (City of New York, 2013), and second, various health department documents, including a community checklist for health sector resilience informed by Hurricane Sandy (Toner et al., 2017). She also noted that the focus throughout her presentation would be on planning and response coordination, not infrastructure. This was not because NYC is not working to make its health care system infrastructure more resilient to a changing climate, but because planning and response are her expertise and an area in which the NYC health department is taking a lead role.

NYC's Health Care System

Like the city itself, Quinn said, NYC's health care system, which serves all of the city's more than 8 million residents, is diverse and complex. She identified four sectors: (1) acute care and psychiatric hospitals, which at the time of Hurricane Sandy included three Veterans Health Administration hospitals; (2) residential providers, which include NYC's 173 licensed nursing homes, 77 licensed adult care facilities, and more than 1,000 other resident programs that provide housing support for people with substance abuse disorders, developmental disabilities, and other behavioral and mental health problems; (3) community-based providers, which Quinn described as any provider in the community that offers ambulatory care services, including nearly 500 federally qualified health centers, more than 100 dialysis centers, 67 methadone maintenance treatment programs, and "countless" other providers located in more than 10,000 buildings across the 5 boroughs; and (4) home-based care providers, which Quinn said was difficult to describe, but the number is rapidly growing and is estimated to have been serving more than 100,000 households at the time of Hurricane Sandy.

The providers and facilities across these four sectors make up an interconnected web of health care services, with an impact to any one service having, Quinn said, a "cascading" effect on the others. To provide a sense of the number of facilities at risk for coastal storm impacts, Quinn mentioned a 2014 analysis showing that NYC has 67 health care facilities in the highest risk flood zones. These facilities include 11 hospitals, 33 nursing homes, and 23 adult care facilities for a total of 15,289 licensed beds.

Quinn asserted that before discussing the NYC health care system experience with Hurricane Sandy, it would be helpful to discuss the city's experience with Hurricane Irene, which preceded Sandy by about 14 months. Irene ended up causing less damage than originally predicted. Nonetheless, she said, its approach prompted an evacuation of health care facilities in advance of the storm, thus providing the city with an opportunity to test its evacuation system. Yet to some, Quinn explained, the risks of an evacuation to patients and staff, combined with the expense of the Irene evacuation operation, seemed to have been unnecessary when it turned out that the storm was not as severe as originally predicted. Ultimately, NYC and New York state used the experience to make some improvements in planning for future pre-storm evacuations. Yet, at the same time, this newly gained first-hand knowledge of the risks of a large-scale evaluation, combined with the ultimate ease of repatriation to undamaged facilities after Irene had passed, influence decision making at many levels as Hurricane Sandy approached the NYC area just over 1 year later. As a result, although lessons were learned across the health care system from responding to Hurricane Irene, Hurricane Sandy proved challenging in unexpected ways.

Evacuations and Other Impacts of Hurricane Sandy on NYC's Health Care System

Hurricane Sandy was a highly unusual storm, Quinn continued to explain. First, it was unusually massive, with a wind field of nearly three times that of Katrina. Plus, it took an unusual course, what Quinn called "the rare westward hook," moving out into the Atlantic before making a sharp left turn and hitting New Jersey squarely on the coast. In addition, it was unfortunately timed, making landfall coincident with a high spring tide due to a full moon. The storm surge in lower Manhattan ultimately reached 14 feet, a new record, which was 30 percent higher than the previous record set in 1960. Finally, the course of the storm pushed water up New York Harbor, leading to two different types of damage—first, wind and wave damage to coastal communities on the Atlantic coast, and second, still water flooding in the upper harbor and water damage to mechanical, electrical, and telecommunication systems.

Quinn suspected that many people can recall images of the health care evacuations that occurred as a result of Hurricane Sandy—patients being carried by hand down dark stairwells illuminated only by the light of cell phones, long lines of ambulances waiting to move patients out of nursing homes and hospitals that were suddenly without generator power, and families searching for loved ones who had been emergency evacuated out of long-term care facilities and transferred to city-run special medical needs shelters.

Although the story of Sandy's impact on the health care system goes beyond health care evacuations and touches on every component of the health care system, health care evacuations played an important role, Quinn continued. In total, 37 NYC health care facilities were evacuated. This included 5 acute care hospitals, 1 psychiatric hospital, 17 nursing homes, and 14 adult care facilities. The health department's count of evacuated persons from these health care facilities was 6,451. Quinn noted, however, that this number does not include patients who were rapidly discharged from hospitals or nursing homes or transferred between or within networks in advance of the storm. She suspected that many more patients were affected. By the department's best estimate, approximately 1,800 of the more than 6,000 persons evacuated were transported to special medical-needs shelters. Additionally, several hospitals set up long-term care units in alternate spaces, such as cafeterias and conference rooms, to accommodate influxes of evacuated long-term care residents transferred by emergency medical services.

Also, as planned, the New York State Department of Health activated the Healthcare Evacuation Center (HEC) to coordinate the evacuation and repatriation of health care facilities. HEC is led by the state health department, Quinn noted, but is supported by key NYC stakeholders, including the city's local health department and emergency management agency, the Greater New York Hospital Association, and others. At the time, she said, HEC had been designed with Hurricane Irene–like conditions in mind, which meant that the HEC leadership and staff had to continually modify their protocols and procedures to accommodate evacuations conducted during and after the storm, as opposed to pre-storm. Plus, Quinn added, the facilities supported by HEC experienced lengthy recovery periods, complicating repatriation.

In addition to the evacuations, Quinn continued, the NYC health care system experienced many other impacts. Ten hospitals remained open and did not evacuate in spite of power outages and flooding. At least 65 long-term care facilities lost power at some point during and after the storm. More than 500 buildings that housed community providers were in inundated areas. Those 500 buildings, Quinn noted, represented 5 percent of all the buildings in NYC that house community providers. Another 1,200 buildings were located in areas that lost power. All provider types, but particularly home-based care providers, were significantly impacted by transportation disruption, travel bans, and lack of access to fuel.

Quinn explained that one result of these varied impacts of Hurricane Sandy was an overall reduction in health care system capacity. In the immediate days after Sandy, city-wide hospital bed capacity was reduced by 8 percent. The same 8 percent reduction was also observed in the long-term care sector in the immediate days after the storm, with about 5 percent fewer long-term care beds available 4 weeks after the storm. Primary care and mental health services were also impacted, especially in hard-hit areas. According to Quinn, a 2014 survey conducted by the city health department indicated that, among 40 primary care providers in the Rockaways, a neighborhood in NYC, 95 percent closed or relocated at some point due to Sandy. Of those, 38 percent were closed for more than a month.

Meanwhile, as the health care system was experiencing an overall reduced capacity, Quinn added, some parts of the system were experiencing a surge in demand for services. Not only did some emergency departments see storm-related increases in emergency department visits, but the NYC health department syndromic surveillance system also identified a surge of visits for basic medical needs, including methadone (for patients in methadone maintenance treatment programs), oxygen, dialysis, and medication refills.

Key Concepts Highlighted by the Hurricane Sandy Experience

Rather than lessons learned, Quinn preferred the term "key concepts," that is, "things that we knew ahead of Sandy, but were just really brought home by the experiences we had during that storm."

The first is that NYC's health care services are interdependent. Thus, even if hospitals have well-developed emergency plans for evacuation and shelter, less-resourced, non-acute, long-term care, and ambulatory services are still vulnerable to disasters. Importantly, Quinn explained, these vulnerabilities have a direct and immediate impact on hospitals.

Second, health care evacuation decision making is complex. Safe evacuation, she observed, takes time. The decision to evacuate must be made more than 72 hours before a storm arrives, that is, during a period of time when the forecast is likely to be uncertain and the weather beautiful. Because the risk of evacuating fragile patients may outweigh the risk of sheltering in place, she said, "the consequences of a decision to evacuate can truly be life or death." She went on to say, "Of course, hindsight is 20/20, so leaders will be criticized for bad outcomes, regardless of whether their decision in the moment was the right one."

A third key concept driven home by the Sandy experience, Quinn continued, is the need for surge spaces of different types. Special medical needs shelters and even hospitals are not designed to accommodate the needs of residents of nursing homes and adult care facilities. Yet, long-term care facilities are typically already operating at 100 percent capacity, with significant barriers to increasing surge capacity in that sector.

A final key concept is that planning for coastal storm and other natural disasters needs to consider both immediate and long-term interruptions to health care services. Following the storm, ambulatory services in some hard-hit areas remained closed for a long period of time, Quinn observed, and some long-term facilities never reopened.

Building a More Resilient Health Care System

Quinn described several ways that these key concepts are being translated into actions to develop a more resilient NYC health care system. First are the strategies outlined in the *A Stronger, More Resilient New York* report (City of New York, 2013). These include ensuring critical providers' operability through redundancy and the prevention of physical damage and reducing barriers to care during and after emergencies.

Next, Quinn remarked on the state health department having made enormous strides in improving HEC operations, specifically in developing an online system for tracking evacuations and transportation requests.

Additionally, the NYC health department has leveraged hospital preparedness programs and public health emergency preparedness funds to support a number of programs designed to improve health care system preparedness. Many of these programs, Quinn noted, focus directly on the long-term care sector. For example, the department has provided nearly all NYC long-term care facilities with radios for redundant communications, and also offers a free technical assistance program for nursing homes and adult care facilities. Since 2015, the city health department has been collaborating with the local hospital association, the Greater New York Hospital Association, to co-lead an interdisciplinary workgroup to address some of the challenges of large-scale patient evacuations. The focus has been on providing guidance for hospitals around standardized patient transfer forms and face sheets (which contain basic patient identifying, admissions, and discharge information), agreeing on definitions for bed types to facilitate the matching of patients to receiving hospitals, and working on plans for advancing access to medical records between sending and receiving facilities.

Resiliency as a Primary Goal

Quinn shared some final thoughts on resiliency and why she thinks a changing climate requires approaching health care system preparedness with resiliency as a primary goal.

First, she noted that when a disaster impacts the health care system, the system is going to break in the places where it is already the weakest. If a long-term care facility "is already squeaking by on razor-thin margins just to stay open on a good day, they are going to have difficulty accommodating a surge of 20 percent of their licensed beds during a storm," she said.

As another example, in a community with few primary care sites, if one or more of those are damaged in a hurricane, then that community, which already had difficulty accessing care, now has no ability to access health care services. For Quinn, this observation means, she said, "health care system preparedness has to go well beyond making plans or even training and exercising those plans. We need to move towards a resiliency framework where our goal is to make our day-to-day system stronger [and one that offers] more consistently high-quality health care services to everyone in our communities." She said that unless that kind of everyday strength and high quality is achieved, it should come as no surprise that a health care system cannot do what it needs to do when an emergency happens.

Although her remarks were based on coastal storms and the need to build more resilient health care systems with the capacity to respond to the more frequent extreme weather events that likely will occur in future years, Quinn stressed that climate change has other effects as well. In her opinion, responses to these other effects, especially as they grow increasingly unpredictable, will also benefit from using a resiliency lens to strengthen health care systems and to ensure more equitable and high-quality health care. As just one example, she cited the threat of emerging infectious diseases. Responding to this threat requires stronger day-to-day infection control practices, consistent adherence to recommended travel screening for patients with possible communicable diseases, and improved communication between health care facilities and public health authorities. As another example, the potential for worsening heat waves makes clear the need to consider adequate housing and access to city services as a public health problem and to ensure that health care facilities can manage a brief heat-related patient surge as easily as they can manage a peak seasonal flu surge.

In conclusion, to ensure the best possible outcomes for all community members, Quinn called for more resilient and adaptable systems and for stronger relationships and better collaboration before, during, and after emergencies.

PLANNING FOR CLIMATE CHANGE IN NEW HAMPSHIRE[2]

Matt Cahillane began by discussing how to start a conversation on climate change and its impact in the northeast without even mentioning the words "climate change." Specifically, he suggested saying either "it is getting warmer" or "it is getting wetter" or mentioning "extreme weather," "sea-level rise," or "storm surge." Not only does not mentioning "climate change" avoid argument, but, in his opinion, the conversations, particularly

---

[2] This section summarizes information presented by Matt Cahillane, New Hampshire Department of Health and Human Services, Concord, New Hampshire.

around extreme events, can become very memorable. He showed a handful of images of past extreme weather events in New Hampshire, including images from the Keene floods in 2005, the 2008 ice storm, Hurricane Irene in 2011, and a southern New Hampshire heat wave in 2015. Using events like these as starting points for a conversation, he explained, allows him to "get in and engage people and talk about things that affected their life in the state."

He then went on to discuss several climate and health projects currently under way in New Hampshire and work that he has been involved with under a CDC BRACE grant.

Climate and Health Projects in New Hampshire

Currently, New Hampshire is involved with four local climate-related public health projects. The first is a heat stress and elders project that has been under way in the Upper Valley, near Dartmouth College, for the past 2 years. The second, in the Lakes Region, is on habitat change, ticks, and tick-borne disease, mostly Lyme disease. The third is a storms and resilience project in the Monadnock area of the state. The fourth is another, newer heat stress study, but in the Greater Nashua area.

Additionally, the state is involved with multiple research studies in climate and health. As just one example, in the Portsmouth area, researchers are testing oysters for *Vibrio* bacteria to determine whether cooling and shading the shellfish can reduce the risk of foodborne disease. He noted that there have been a few recent *Vibrio* outbreaks up and down the eastern seaboard and that more should be expected if the water temperature in the ocean continues to rise.

In addition to these public health projects, the state is also conducting two state-level studies out of its main state offices in Concord. One is a weather and injury study at Plymouth State University; the other is a study on heat stress and emergency department visits.

The Building Resilience Against Climate Effects Framework

CDC's BRACE framework has been very helpful for New Hampshire's state-level work, Cahillane continued (see Figure 5-1), and he and colleagues are hoping to adapt the framework at a local level as well. He described the framework as a five-phase public health improvement cycle. The first step is the forecasting of climate impacts and the assessing of vulnerabilities. For New Hampshire, this involves working with the state's academic institutions to downscale data from large global models so that they are applicable to New Hampshire, which has its own topography, and predicting how climate is going to change in different regions of the state.

FIGURE 5-1 BRACE (Building Resilience Against Climate Effects) framework.
SOURCES: Cahillane presentation, March 13, 2017; CDC, 2017.

They are trying to predict, for example, how many days will have temperatures over 95°F and how many precipitation events will involve one or more inches. These forecasts and assessments, Cahillane said, help local communities to better understand the changes for which they will have to prepare.

The second phase of the BRACE framework is projecting disease burden. This is probably the most challenging component of the process, Cahillane said. The state health department does more retrospective work than it does projections into the future, he explained. However, it does a bit of projecting work with aging and demographics. Plus, now it is attempting to predict the effect of extended pollen seasons on asthma and allergies and determine whether the effects can be quantified.

Third is the assessment of public health interventions and identification of evidence-based public health practices with known impacts on current problems and perhaps future problems as well. For example, Cahillane asked, is there a good intervention for asthma? The answer, he said, is "yes." He referred to the CDC *Guide to Community Preventive Services* (CDC, 2017). But for things like weather-related slips and falls, heat stress, and other similar issues, there is not enough information, he noted. He remarked that the New Hampshire state public health office is working to build that evidence base.

The fourth phase of the framework is developing and implementing a climate change adaptation plan, which Cahillane described as "basically, putting together a plan that looks like it is going to work well, that reflects a community's needs and resources, and that is feasible for that same population."

The final phase of the BRACE framework is evaluation. In other words, Cahillane asked, did what was implemented actually work? Can it be reproduced at the local level? Is the evidence base being built? Can performance be improved by understanding what has happened?

Cahillane went on to discuss in more detail how each of these five steps is being implemented in New Hampshire.

Forecasting Impacts and Assessing Vulnerabilities

In addition to working with local universities and other modeling experts to better understand some of the more complex downscaling of global data, the state health department also relies on CDC guidance to assess vulnerabilities (CDC, 2014). Additionally, they use existing tools, such as the New Hampshire Social Vulnerability Index (SVI), and websites with regional hazard data. The New Hampshire SVI, Cahillane explained, is based on Census data and is a composite indicator of more than 14 metrics, such as education, income, disability, age, and whether someone is living alone (see Figure 5-2). When a disaster hits a community, regardless of whether the disaster is natural or not, the vulnerability index provides a sense of how well a community will likely be able to adapt to the changes and helps planning at the state and, to a lesser extent, the local level.

Projecting Local Health Burdens

Cahillane reiterated the challenge of this step of the BRACE framework. One of the existing tools that the state public health department uses to assess health trends is its dynamic web portal, NH WISDOM.[3] By plugging in health outcomes or behavioral issues that they want to examine, users can view county- and city-level trends. While the portal does not provide community-level details yet, Cahillane explained it is helpful because users can determine immediately whether, for example, slips and falls are higher in some regions (e.g., as they are in the greater Monadnock area) or

---

[3] See https://wisdome.dhhs.nh.gov (accessed March 9, 2017).

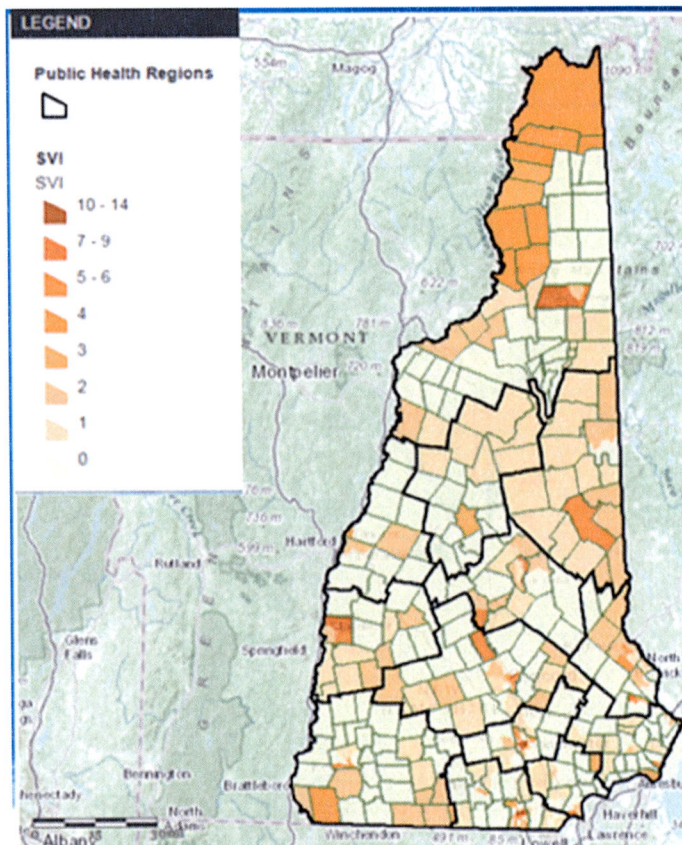

FIGURE 5-2 New Hampshire Social Vulnerability Index (SVI).
NOTE: SVI scores for public health regions across New Hampshire; scores are composite indicators of more than 14 socioeconomic, demographic, and housing and transportation metrics.
SOURCE: Cahillane presentation, March 13, 2017; reprinted with permission from Cahillane; New Hampshire Department of Health and Human Services, 2017.

lower in others (e.g., as they are in greater Nashua) compared with average rates across the state.

State-wide priority health issues that have risen to the top as a result of this work, Cahillane continued, include heat stress; habitat change (with an impact on ticks and Lyme disease); and extreme weather-related injury and resilience. Priority areas in development include the effect of rising water temperatures on bacteria (with implications for shellfish safety) and changing precipitation and watershed-related health outcomes (e.g., the health impacts of the weather becoming wetter, or drier, depending on the season).

Assessing Public Health Interventions

Based on these priority health issues, the state health department has been working with local communities to define exposure pathways. For example, with Lyme disease, Cahillane asked, "where are the places that an intervention at the individual level would break exposure?" This could include helping individuals to avoid tick habitats, helping people to apply repellant, or helping people to learn to do tick checks. He said, "We are always looking for a place where we can break that cycle."

In addition to defining exposure pathways, the state is also reviewing existing evidence, for example, by conducting literature reviews. But conducting these reviews at the local level is a challenge, Cahillane explained, because local health departments often do not have epidemiologists who can do this work.

Meanwhile, the state is also continually building its evidence base, for example, through work conducted as part of the New England Heat Study. In cooperation with Maine and Rhode Island, New Hampshire is trying to pull data together to evaluate temperature changes over time and their effects. He noted that this is a retrospective study in that, for example, they are examining the effects of past warmest days, not just with respect to how many people showed up in the emergency department because they had dehydration, heat stress, or an elevated temperature, but also with respect to the wide range of moderate impacts associated with heat. Cahillane noted that the number of people with respiratory, cardiovascular, or kidney disease who show up in the emergency department during periods of high temperature is actually larger than the number of people who show up for heat-related injuries—information the health department is working to share with communities.

Finally, with respect to the use of existing tools for assessing interventions, Cahillane mentioned again CDC's *The Guide to Community Preventive Services*. Additionally, he mentioned the CDC Communities of Practice effort at the national level that is encouraging states to dig into the literature and understand it better.

Developing a State Strategic Plan

State-level strategic planning involves identifying ways to build workforce capacity, improving education and outreach, and improving policies. Cahillane stressed that it is not just legislation that needs to be addressed, but also policies at the local level. This could include, for example, something as simple as deciding whether a daycare center should have an opt-in or opt-out policy for using repellants. Currently, he said, the centers have an opt-in policy, whereby parents who want repellants applied to their children must sign a form. But if repellants (i.e., exposure to chemicals in the repellants) are less risky than contracting Lyme disease, having an opt-out policy would be more effective, so parents who do *not* want repellants applied can opt out.

Cahillane reminded the workshop audience of the four state-level public health intervention projects currently under way in New Hampshire and then discussed in more detail the two projects that are a little further along. First, the Upper Valley Heat Project in the Dartmouth region has, thus far, involved engaging with local advisors to assess wants, needs, and hazards; hiring a regional planning commission to conduct the assessment; prioritizing areas of rising temperatures, heat stress, and injury; targeting older adults, particularly those living in the community (i.e., as opposed to living in nursing homes) and shut-ins; intervening via education lectures to caregivers and older adults on risk factors; and evaluating the results with a before and after test of caregivers' skills in heat risk factors.

The New Hampshire Lakes Region Tick Project has taken a similar approach, Cahillane said, by engaging the local public health advisory council; prioritizing habitat change, tick exposure, and tick-borne disease (e.g., Lyme disease); targeting outdoor youth in recreation programs; intervening via education of adult recreation staff and students to decrease the risk and reduce tick exposure; and evaluating results via pre- and post-testing of staff for risky and protective behaviors.

Evaluating Impact

The evaluations of the Upper Valley Heat Project and New Hampshire Lakes Region Tick Project, Cahillane continued, have revealed a limited ability in the early stages to show quantitative change in knowledge or behaviors via pre- and post-testing. However, both projects have successfully shown some qualitative change in the amount of community collaboration among the local planning commission, recreation department, and elder care services. Additionally, Cahillane suggested that now there may be a stronger case for making changes to policy and rules via engagement with decision-makers, particularly with respect to risks for moderate health impacts. For example, through engagement with the National Weather Service, state efforts have encouraged a lowering of the threshold for when heat warnings are issued.

Lessons Learned

Cahillane concluded by listing some lessons learned. First, a standard planning process, such as the CDC BRACE framework, is worth more than the actual final written plan, which, he said, "often ends up on [the] shelf." Plus, the framework supports greater collaboration.

Second, engaging with opponents of climate science is rarely successful for the community good. Cahillane said that, based on his experience, many people want to engage in the climate debate because they want to make other people uncomfortable and decrease civil discourse. He quoted Bernard Shaw, "I learned later in life not to wrestle with a pig. You both get dirty, and the pig enjoys it." Cahillane explained that his intention in quoting Shaw was not to call anyone a pig, rather to emphasize that some people enjoy arguments not because they want to actually engage in discussing what might be an uncomfortable issue, but because they want everybody to feel "dirty" afterward. "If you can get everybody in the room feeling uncomfortable," he said, "you can suppress that conversation."

A third and final lesson learned is that a seed grant project led by a local health agency takes a lot of time, patience, and support. Cahillane estimated that getting the four state-level public health intervention projects up and running over the past couple of years has taken more than half his time and that each project has required about $60,000 to $65,000 thus far (i.e., about $20,000 per year per project from the state, plus other support).

CITY OF BALTIMORE:
COMMUNITY RESILIENCE IN A CHANGING CLIMATE[4]

Kristin Baja began by emphasizing that some populations are more vulnerable to the impacts of climate change than others, and thus, it is important to consider equity in climate planning work. Indeed, the City of Baltimore relies on the use of equity as a lens for thinking about climate change and climate planning. Baja then went on to discuss in detail various components of Baltimore's resilience-based approach to climate planning. First, however, she noted that the city also faces unique climate challenges, stemming from its location. Specifically, Baltimore is located at the top of the Chesapeake Bay and at the base of four major watersheds. "So we basically have water coming from everywhere," Baja said. At the same time, the city is slowly "sinking" into the bay, she said.

---

[4] This section summarizes information presented by Kristin Baja, Office of Sustainability, City of Baltimore, Maryland.

Baltimore's Unique Approach to Climate Planning

Baltimore's unique approach to climate planning is based on using the federal government's required All Hazard Mitigation Plan and incorporating climate change into that plan. Baja described the city's incorporation of climate change into the mitigation plan as "going a little bit against" what is expected in that it acknowledges climate change.

Moreover, not only is Baltimore acknowledging climate change, but the city is using "climate change" in all of its messaging. However, Baja explained, some of the specific language used changes depending on who is participating in the conversation. For example, she and colleagues usually talk about disaster preparedness when talking with residents or business owners. But when talking with their funding community or private partners, they talk about resiliency. She defined resiliency as: "The ability of our community to anticipate, accommodate, and *positively adapt to or thrive* amidst changing climate conditions or hazard events and enhance quality of life, reliable systems, economic vitality, and conservation of resources for present and future generations." Baja highlighted "positively adapt to or thrive" as a key component of what Baltimore is trying to do with all of its climate planning.

In addition to its acknowledgment of climate change and focus on resiliency, the city also does something a little different, Baja added, by considering not just climate-related "shocks," but also "stresses" (see Figure 5-3). Shocks are typically single-event disasters, such as fires, hurricanes, and floods. Stresses, while they can be exacerbated by shocks, are factors that pressure Baltimore on a daily or recurring basis, such as endemic violence and high unemployment. The images of recent Baltimore shocks that are shown in Figure 5-3 are ones that Baja captured on her iPhone. They include a collapsed retaining wall, people attempting to drive through 6 inches of rain that had fallen in just a few hours, and a severe storm known as a *derecho*. The derecho hit in 2012, knocking out power for about 7 days during one of Baltimore's highest heat waves (i.e., 95°F and higher for that period of time). While shocks like these are expected to become more severe and more intense in the future, in Baja's opinion, what is being left out of many conversations are the stressors.

Talking about stressors requires acknowledging the past, Baja explained. In Baltimore, it means acknowledging historic planning practices that have been intentionally racist and have segregated populations. For Baja, it means also acknowledging that she, herself, is a privileged individual who did not experience the racist and segregationist impacts of those historic planning practices. Therefore, when she walks into a room when working as the city's Climate and Resilience Planner, the first thing she does is acknowledge her privilege and then ask if people are willing to

FIGURE 5-3 Shocks and stresses affecting the City of Baltimore. Recent climate-related Baltimore "shocks" (i.e., single-event disasters) versus "stresses" (i.e., factors that pressure Baltimore on a daily or recurring basis).
SOURCE: Baja presentation, March 13, 2017; (top left) reprinted with permission from Baja; Baja, 2016b; (top right) reprinted with permission from Lasinski; Lasinski, 2016; (bottom) reprinted with permission from Baja; Baja, 2016a.

sit down and have a conversation with her. She tells them that she wants to listen, hear their stories, and learn about displacement and what it has meant for their commutes and daily lives. (Baja noted that, from 1951 to 1971, 80 to 90 percent of the 25,000 Baltimore families displaced to build new highways, schools, and housing were Black.)

Baja described what is known as Baltimore's "Black butterfly" and "white L," with the Black butterfly representing where African Americans have been "told" they can live since the 1910 ordinance segregating Blacks and whites by block. The white L is where most of the city's transportation, as well as the Inner Harbor, are located. Although she did not describe its details, Baja referred to a document put together by Lawrence Brown that lists the advantages in the white L versus the Black butterfly regarding certain policies and practices (Brown, 2016). Additionally, she showed maps illustrating the socioeconomic differences between the Black butterfly and white L, with low income and unemployment being highly prevalent in the Black butterfly area of the city (see Figure 5-4).

Baja emphasized the importance of not just treating everyone equitably (i.e., by providing the necessary support to ensure that everyone has the same access), but removing systemic barriers that cause the inequity in the first place (i.e., certain historic practices and policies).

Equity as a Lens

Using equity as a lens, Baja explained, involves first prioritizing neighborhoods that have the highest vulnerability and historic disinvestment. Next, it also involves actively listening to and engaging with residents. For example, she and her colleagues stay away from PowerPoint and other presentations. They focus instead on interaction. Using equity as a lens, Baja continued, also involves thinking about new opportunities in all of the

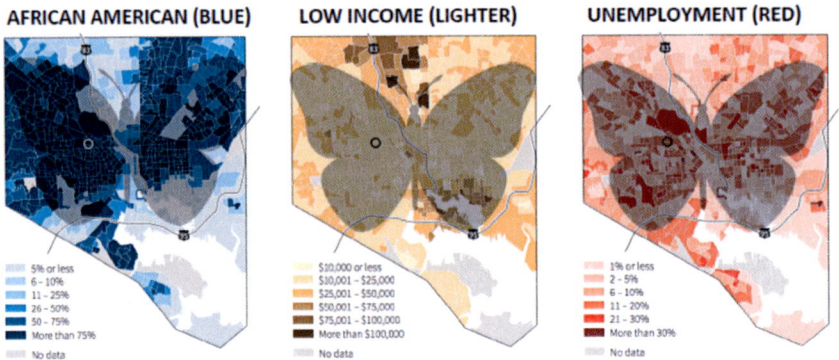

FIGURE 5-4 Maps illustrating Baltimore's socioeconomic differences: (from left to right) the "Black butterfly" and the "White L" (referring to the shapes that the darker and lighter spots on the map create) overlaid by a map of the African-American population, an income map, and an unemployment map.
SOURCES: Baja presentation, March 13, 2017; Brown, 2016; adapted and reprinted with permission from Reuters; Reuters, 2015.

city's different initiatives. Rather than telling people about an infrastructure change, such as a new park, they consider the "whole system" (e.g., job training, green job opportunities). Finally, they try to make sure that individuals who have been overlooked or do not have the resources to get to meetings are being reached, for example, by encouraging residents to choose meeting locations in places that they trust, such as a home, a house of worship, or a community center.

At the actual meetings, Baja and colleagues usually bring in some element that is both interactive and fun so there is additional incentive to be there. They also bring food and offer free childcare, and they try to provide transportation for those who do not have access. She showed images from one meeting, where a 24-foot-long wall of paper had been constructed to illustrate historic racism and practices in Baltimore as they relate to sustainability and resilience, including where changes have occurred and where there are issues that still need to be addressed. At the meetings, Baja and colleagues also ask residents about their own definitions of equity and sustainability, what residents are going to do, and where residents see themselves being part of the city's efforts. Baja said, "We are looking for implementation partners, not just for feedback."

Implementation

Baja expressed pride in the fact that Baltimore's Disaster Preparedness Plan "doesn't sit on the shelf." Baja herself uses the plan on a daily basis. It has 231 actions around resiliency. The plan takes a whole-block approach, she explained, which involves thinking about not just the built infrastructure of a neighborhood, but also other opportunities with natural systems (e.g., trees and greening, storm water, heat sensors) and public services (e.g., cool roofs, weatherization, energy education, renewable energy).

An initial, immediate focus of the work that Baja has been involved with in Baltimore, and a focus that Baja perceived as being key to initiating many other city programs, was enhancing community preparedness. They help communities to make plans, build kits, and help each other. The make-a-plan component of the work involves sitting with residents and helping them create an emergency plan and identify their evacuation and sheltering opportunities. The planners also work together with residents to build the emergency kits, rather than the city handing out kits and saying, "Here is your emergency kit." Baja explained that they talk about every single item that is in the kit, such as how to hand-crank the radio and how to use the solar panel to charge phones. Finally, the help-each-other component of community preparedness is, in Baja's opinion, the most important piece. It involves doing a lot of asset mapping with community members to identify what exists in the community and what is missing.

Baja went on to discuss several of the community disaster preparedness activities under way, starting with help and safe signs. She described
Baltimore as a city of row houses, with tiny streets and many houses
packed into those streets. To help the community emergency response
teams, as well as first responders, in this environment in the event of an
emergency, the city has developed help and safe signs that can be placed
in windows indicating whether an occupant needs additional assistance.
Baja recalled how they were able to distribute signs to many, but not all,
of the 265 Baltimore neighborhoods before the 2015 riots and that many
households where the city had conducted outreach did in fact use them
during those riots.

Residents have identified several missing pieces in the community preparedness work, Baja continued. For example, many residents do not have
access to opportunities to evacuate because they do not have vehicles. Additionally, many residents are not setting aside extra food or water because
they are dealing with multiple jobs, childcare, or other pressures on a day-
to-day basis that keep them from proactively preparing. The largest missing
piece, though, Baja said, is that many residents do not have a safe place to
go where they can find trusted information.

In response to the need for safe places with trusted information, the city
is currently piloting the concept of resiliency hubs. Baja was unaware of any
other city in the United States that has adopted this concept, although she
noted that Los Angeles has taken an interest. She defined a resiliency hub
as consisting of "a building or set of buildings and neighboring outdoor
space that will provide shelter, backup electricity, access to fresh water, and
access to resources such as food, ice, charging stations, etc., in the event
of an emergency." Baja and colleagues are working with members of communities to identify trusted neighborhood sites that are already being used
daily, such as houses of worship or community centers, that could serve as
resiliency hubs. Once these sites are identified, the program then provides
funding to enhance or make the site more resilient, for example, by installing solar panels with battery back-up and by providing materials for storing
food and water. Thus far (i.e., at the time of this workshop), four pilot hubs
are already operational and another two are potentially opening within the
next year. The hubs are also being used by other city programs, including
health department programs.

In addition to the help and safe signs and resiliency hubs, another
specific community-level tool being implemented as part of Baltimore's
disaster preparedness efforts is its ambassador program, a peer-to-peer engagement network of more than 150 ambassadors who help with outreach.
The program has divided the city into 10 districts with similar population
sizes and assigned a team leader to each district. Each district also has a
lead ambassador and several community ambassadors. The team leader

and ambassadors have gone through several trainings on equity and sustainability over the past year and have been provided with toolkits that have surveys, draft presentations, and other tools. Baja remarked that the ambassadors themselves help to design these toolkits, which she said has been "great" because their involvement ensures that they receive the tools they need.

Another key piece of the city's disaster preparedness efforts, Baja continued, is that instead of just thanking community members for telling their stories and providing information, she and her colleagues in the city office show how they are actually using the information. One way they do this is through Baltimore's Every Story Counts campaign, which Baja described as a communications campaign "about the amazing people that live in Baltimore that are doing things on their own without support." She showed an image of a group of women who help immigrants and non-native English speakers to find resources, health care, and other opportunities within the system.

Additionally, the city makes use of Turtle, a character who Baja and colleagues thought would be a fun way to talk about climate and resiliency with kids. As it turns out, Baja said, adults like Turtle just as much as the kids do. A turtle was chosen not only because turtles are native to Maryland, but also because they carry their homes on their backs and are adaptable to both land and water. Thus, Baja said, they serve as a good image for messaging about preparedness.

Baja and her city colleagues also use a lot of games in their trainings. As an example, she mentioned the "Game of Floods," which was developed through the Urban Sustainability Directors Network and that the city is adapting for use specifically in Baltimore. The game is actually a vulnerability assessment, Baja noted, but is much more fun and engaging than a typical vulnerability assessment. Players pick a role and then identify how much money and resources are available for that role.

Finally, Baja commented on the many other Baltimore initiatives that her office integrates into its own work, for example, via the resiliency hubs. An example is Vacants to Values, a strategic demolition and vacant lot revitalization program. Baja said, "We are seeing vacant lots as opportunities, rather than something that is negative."

In conclusion, Baja mentioned that she and her colleagues are trying to measure the impacts of many of these programs as they go along. For example, they have installed 250 heat island sensors in the communities to see if any of the retrofitting they have done to the built environment is having an impact. They are also conducting trainings for community members not just in emergency response, but also in tree care and planting and weed removal.

DISCUSSION

Following Baja's presentation, she, Quinn, and Cahillane engaged in an open discussion with the audience. Some of the questions were directed at specific state-wide or city-wide climate planning programs (i.e., NYC, Baltimore, or New Hampshire), while others were addressed to all panelists (e.g., funding for their respective programs).

The NYC Health Care System: Collaboration Across Sectors

Moderator Paul Biedrzycki asked Quinn how the four sectors of NYC health are interlinked in other ways in addition to preparing and responding to hurricanes and other weather-related emergencies. Quinn replied that she would be remiss if, in talking about NYC's health care preparedness efforts, she did not talk about health care coalitions, which she said have been an important national initiative. NYC has several types of such coalitions, with about 23 of them coalescing into what she described as a "coalition of coalitions," through which all four sectors of the health care system are represented. Additionally, she mentioned increasing efforts to conduct planning exercises and trainings across sectors and the success they have had over the past couple of years bringing health care partners into health department exercises. For example, in 2016, the NYC health department solicited help from a couple of representatives from large health care systems in the planning of a cyber-attack exercise to ensure that the exercise was as realistic as possible. The representatives also participated as observers in the actual exercise.

When the panelists were asked by an unidentified audience member whether any of the communities or agencies conducting emergency preparedness are more likely to also be coordinating health care delivery more generally, Quinn replied that it "goes both ways." Many changes are under way in the health care delivery system, with many people being taken care of at lower levels of care. This is impacting how hospitals think about emergency management, she said, or at least "it should be" changing how they think about their emergency management programs. Her hope is that, at the same time, all of this emergency preparedness work is also impacted by day-to-day service delivery. "It is my belief," she said, "that if we are planning for things that are only going to be pulled out during an emergency, we are, first of all, losing an opportunity to improve the health care system that really needs a lot of improvement. But we are also designing systems that are not likely to work in an emergency because they are not being used day to day."

As an example, she mentioned an NYC health department project being done in collaboration with the Greater New York Hospital Association to develop improved patient transfer forms. She and her colleagues sat down with a broad group of stakeholders from different health care systems and transfer centers, including physicians, nurses, and emergency managers, to agree on a set of fields of critical information that are necessary for taking care of a patient who is being rapidly transferred during an emergency. The form has since been disseminated to all hospitals, Quinn said, and all hospitals have been asked to implement it as their day-to-day transfer form. In fact, some of the hospital systems have started working with their electronic health record vendors so that simply pushing a button pulls those fields from a patient's electronic health record. Once this new form is implemented city-wide, Quinn said, the city will have a system whereby all hospitals are using the same information to transfer patients. Not only will transfers be safer, they will also be quicker, improving hospital through-put and making the form a win–win improvement. "We want to see our preparedness program more focused on those day-to-day improvements," she said.

Hurricane Sandy Evacuation

Quinn was asked by an unidentified audience member whether any data were collected on the impact of the Hurricane Sandy evacuation, with such a sharp disruption in the lives of the estimated 6,000 persons evacuated from health care facilities. For example, were any data collected on mortality or other outcomes? Quinn answered that the estimate did not account for patients who were quickly discharged or transferred before the evacuation order. To her knowledge, no such data were collected.

Alert Fatigue

Sanne Magnan asked the panelists whether, as storms increase, they expect to see any kind of alert or disaster fatigue and if their agencies will work with communities around this fatigue. "I know that people in our agency have disaster response fatigue," Quinn replied, "myself included." But there are benefits, she remarked. Having frequent emergencies allows for practicing emergency responses and improving those responses. However, she said, as she had described with Hurricane Irene, "you sometimes take away the wrong lessons." Also, she said, there is the "boy who cried wolf phenomenon," which she suspected people in NYC were feeling right "now" (i.e., on the day of this workshop), in light of the "big storm" being predicted for "tomorrow" (i.e., the day after the workshop). In sum, she said, "Yes, we are contemplating a future where we are activated more continuously and trying to figure out how to adapt our systems to meet that challenge."

Cahillane commented that he and others in northern New England are working with the National Weather Service on that exact issue. Fortunately, he said, northern New England is protected by a temperate environment where it does not get that hot, yet new public health data show that even moderate heat can impact people as much as extreme heat. He asked, "So how is that going to change the number of alerts we do per year?" It used to be that only one or two alerts a year were put out when the temperature rose above 100°F. Lowering that temperature to 95°F would increase the number of alerts per year to about four to eight. The number of alerts is not the only concern, in Cahillane's opinion; another concern is the way meteorologists communicate to people, nearly every day, he said, "about something else that is going on." He said, "People are getting hit with a lot all the time."

Baja added that in Baltimore, they try to be strategic with their alerts by differentiating among audiences and identifying those who are most vulnerable. For example, people living in flood plains have already experienced massive impacts. The city has thus established with them a categorical rule to evacuate regardless, given the risk of flash flooding. She described Baltimore's flood plains as "some of the most flood-prone waterways in the country," with some areas experiencing a 20-foot rise in less than an hour. "Because those areas have been flooded out in the past, they listen," she said. But it is different with residents, which is why she and her office are placing such an emphasis on making the resiliency hubs fun locations, with games, where people already congregate and spend time.

Baltimore:
Incorporating Climate Planning into the All-Hazard Mitigation Plan

Terry Allan of Cuyahoga County Board of Health, Ohio, asked Baja to elaborate on the City of Baltimore's approach to community-level emergency management. Allan suggested that other communities could emulate that approach. Baja responded that the work began in 2012, when her office decided they wanted to be proactive, rather than reactive, because they knew they were going to "keep on getting hit with these events." They did it as part of their all-hazard mitigation plan, she explained, because the methodology used in that plan is almost exactly the same as that of a climate adaptation plan. She called it a "very natural overlap." That is, first, they assessed hazards and vulnerabilities based on historic hazards, then they identified how they were going to mitigate those hazards. Then, they approached both their Maryland emergency management agency and Federal Emergency Management Agency Region 3 representatives and asked if they would be willing to allow this approach. Baja said, "We are really lucky to have some good support at both the state and federal levels at the same time."

Now, she said, climate change is 100 percent integrated into the all-hazard mitigation plan. Every action in the plan encompasses climate scenarios, allowing the city to integrate climate change into other big initiatives (e.g., an initiative related to flood plain management). She added that, in her opinion, as climate science changes and as climate hazards occur more frequently, the more these documents should be updated to plan for climate scenarios.

When asked by Jonathan Patz about the importance of optimizing both preparedness and adaptive capacity, Baja said, "We can't just do actions on mitigation anymore. They have to have co-benefits for adaptation."

Patz also asked Baja if any of the community initiatives that her office has been developing are paying off in ways that had not been expected. She mentioned some work they have done with the Waterfront Partnership, where they installed a solar-powered water wheel, "Mr. Trash Wheel," that collects trash from Jones Falls and helps to keep the Inner Harbor cleaner. A couple of added benefits of the wheel are that it keeps boat-clogging agents out of the harbor and that it provides educational opportunities about the use of solar power. As another example, their flood plain work has also provided both educational (i.e., for the real estate community) and insurance (i.e., reduced insurance rates) co-benefits.

Funding

The panelists were asked by an unidentified audience member where they find funding for the different projects and programs they described. Baja replied that Baltimore has been able to secure funding from the Maryland Department of Natural Resources and the National Oceanic and Atmospheric Administration Office for Coastal Management. But the "really cool stuff" they do, she said, such as Turtle and the Every Story Counts communication campaign, is funded through outreach and philanthropy.

In New Hampshire, CDC provides most of the seed money for the projects Cahillane described, he said. "Without a doubt," he said, "it is all the supporters in Congress and the administration who believe in this work and want to get it done." Additionally, he underscored the importance of the support of people in the community who want to engage and who give in-kind and work extra hours.

Scaling

Raymond Baxter, a health adviser, observed that all of the programs described by the panelists embody comprehensive planning approaches.

He asked the panelists about the extent to which their respective programs had been scaled up from something that someone else has already done. Additionally, he asked what it would take to then scale up, to a regional, state, or national level, some of the ingredients contributing to the success described by the panelists.

Baja replied that in Baltimore, rather than scaling up, in fact, local officials are scaling down from the larger, national climate datasets. She also emphasized the usefulness of finding trusted contacts in other places with similar histories to share information, such as which toolkits are most useful, and identifying opportunities for the integration of climate planning into other plans. For example, like Baltimore, Milwaukee also is one of the most segregated cities in the country. Trusted experts can be useful too, she remarked, that is, experts who can help to translate information from the scientific and academic communities.

The NYC health department has been working with the NYC health care system since the beginning of the CDC-funded Hospital Preparedness Program in the early 2000s, Quinn commented. Much of that work, she said, has focused on the acute care sector. It has allowed the department to build strong relationships with emergency managers in all of the NYC hospitals. Only since 2013 have they prioritized expanding those relationships and reaching other sectors of the health care system. She said they have seen some success with long-term care, but are still challenged by urgent care, which she said is a rapidly growing and hard-to-enumerate sector of the system. But her greatest worry, she said, in terms of vulnerability to future emergencies, especially coastal storms, is with home-based care. In terms of scaling up even further what NYC has done, in her opinion, that will require strengthening more of the connections among sectors and then bringing that strength to their regional partners across the state, in New Jersey, and throughout the rest of the area.

For Cahillane, the question is, "How do we ensure that the tax dollars we use and the things we do in state government aren't a waste?" With respect to the BRACE framework he discussed during his presentation, the intervention assessment phase is the one phase where the expectation is that those conducting the assessment will go and find out whether anyone else has implemented the intervention and, if so, if it has been implemented in a similar population. "Let's gather the good work that has been done already," he said, "before we jump forward and put it to work."

Baja added that rather than making more toolkits for local governments, what is needed is "somebody to identify what toolkits actually work." In response, Pamela Russo mentioned Alonzo Plough's work in Los Angeles, where he designed a study to test a toolkit in eight different regions of the city, two neighborhoods per region, with some of the neighborhoods receiving standard of care and the others a resilience training for preparedness (i.e., resiliency to either a hurricane or earthquake or to daily economic or violent stressors). Different components of the evaluation have been published, according to Russo, with respect to which aspects of the toolkit worked and how they refined the kits. The work includes both qualitative and quantitative results.

6

Regional Perspectives from the West

In the final regional perspective panel of the workshop, moderated by Lynn Goldman,[1] Kathy Gerwig provided a perspective from Kaiser Permanente, a large integrated health system with facilities located primarily in California and other western states. Next, Renata Brillinger presented a perspective from the California Climate and Agriculture Network (CalCAN), a coalition of sustainable and organic agriculture organizations that works very closely with farmers and ranchers. The third and final panelist, Fletcher Wilkinson, presented a perspective from the Institute for Tribal Environmental Professionals (ITEP), a 25-year-old organization based in Arizona that works with tribes on a broad range of environmental issues, including climate change. This chapter summarizes these three panelists' presentations. Key points made by the panelists are presented in Box 6-1.

In the open panel discussion with the workshop audience, topics covered included working with vulnerable communities, incentives for innovative approaches to agriculture, Kaiser Permanente and the political climate in California, the challenges of communicating about climate change with the agricultural sector and with legislature, learning from tribes, measuring greenhouse gas emissions in terms of health, and reasons for hope. The discussions around each of these topics are summarized at the end of this chapter.

---

[1] John Bolduc was scheduled to moderate, but had to leave the workshop early due to the coming storm.

> **BOX 6-1**
> **Key Points Made by Individual Speakers**
>
> - In addition to being powerful community forces, hospitals also provide an opportunity, via Community Health Needs Assessments, to identify local health needs and to engage with community members in discussions about ways to meet these needs. Using such assessments, in 2016, 14 of 39 Kaiser Permanente hospitals identified climate implications for health as a priority health need for the first time. (Gerwig)
> - Climate action is Kaiser Permanente's number one 2025 environmental stewardship goal, specifically to be net carbon positive by 2025. Currently, half the electricity that Kaiser Permanente uses to power its California facilities is renewable—mostly solar or wind. (Gerwig)
> - Greenhouse gas emissions are one of many environmental challenges related to agriculture, with an estimated 7 percent of California's emissions attributed to on-farm emissions (i.e., not including upstream or downstream emissions). (Brillinger)
> - California has adopted a highly ambitious set of climate policies, with a vision to reduce greenhouse gas emissions to 40 percent below 1990 levels by 2030. In addition to the state's AB 32 cap-and-trade program, California has begun or is in the process of beginning several new "climate smart" agricultural programs that allocate grants or payments to farmers in return for taking specific steps to protect farmland, reduce water use, capture and convert methane into bio-gas, or build soil carbon. (Brillinger)
> - The overarching goal of the Institute for Tribal Environmental Professionals' (ITEP's) Climate Change Program is to help tribes build capacity to deal with climate change as sovereign nations. ITEP does this mostly through trainings, but also through the provision of information and a tribe-specific toolkit for developing a climate change adaptation plan. (Wilkinson)
> - Because tribal identities are so intimately connected to the places where tribes live, tribal planning is different from a city or state's planning efforts. This sense of place, including the use of traditional foods obtained through subsistence hunting, gathering, and agriculture, needs to be integrated into the entire planning process. (Wilkinson)

CLIMATE + COMMUNITY HEALTH + HEALTH CARE IN THE WEST[2]

Gerwig began by remarking that she would be providing a perspective from Kaiser Permanente, a large integrated health system with medical centers, Permanente Medical Groups, and an insurance plan. Kaiser

---

[2] This section summarizes information presented by Kathy Gerwig, vice president of employee safety, health, and wellness and environmental stewardship officer, Kaiser Permanente, Oakland, California.

Permanente's climate action, she said, is a key pillar in the system's environmental stewardship efforts, which are a component of its community benefit work.

Kaiser Permanente has eight regions, five of which are in the west. Of the health plan's 11.7 million members, nearly 10 million live in western states. An even greater percentage of Kaiser Permanente's greenhouse gas emissions come from the western part of the organization, because that is where all its hospitals are located, Gerwig explained. In the eastern regions, Kaiser Permanente has medical facilities, but not hospitals. Most of Kaiser Permanente's 70 million square feet of real estate are in the western states, a large portion of it in California.

Gerwig commented on the fact that states where Kaiser Permanente facilities are located trend a little differently from the rest of the nation with respect to political context and, thus, opportunities for future work. In the 115th U.S. Congress, the House of Representatives has a Republican majority (238 Republicans, 198 Democrats). But among Kaiser Permanente states, Democrats outnumber Republicans two to one. In the 115th U.S. Senate, again, Republicans hold the majority, but Democrats outnumber Republicans by about four to one among states where Kaiser Permanente is located. Additionally, all of Kaiser Permanente's western states have Democratic governors (California, Hawaii, Oregon, Washington)—an important consideration as Kaiser Permanente's engagements in climate action are "a key pillar of its environmental stewardship."

Health Care Organizations Are Powerful Community Forces

Health care organizations "wear a lot of different hats," Gerwig continued. In addition to clinical care, health care organizations, particularly hospitals, are big emitters of greenhouse gas emissions, big water users, and big waste generators. "They leave," Gerwig said, a "big environmental footprint." Additionally, health care organizations provide many jobs, with hospitals often being the largest employer in their areas. Kaiser Permanente is the largest private employer in California. Thus, they are a "powerful force," she said, for "community work." Plus, they are big purchasers, and they have big supply chains. Whether their supply chains are global or local can have climate implications. Health care organizations are also big landowners, with, again, Kaiser Permanente owning 70 million square feet of real estate. Like most not-for-profit systems, Kaiser Permanente is also a large grant-maker. Finally, health care organizations are big investors. Kaiser Permanente has foundation funds to invest, Gerwig noted. The degree to which an organization's investments are guided by social and climate criteria are important, she remarked.

Hospitals' Community Health Needs Assessments

In addition to health care organizations impacting their communities in the many ways listed above, they also provide an important opportunity via Community Health Needs Assessments. These assessments, which are tied to the Patient Protection and Affordable Care Act, are required every 3 years from all not-for-profit hospitals. Gerwig noted that while the future of this requirement is uncertain, early signals indicate that they have not been targeted to be removed and that they will continue to be an opportunity. She explained that the opportunity provided by these assessments comes from the increased accountability and transparency for how not-for-profit hospitals spend their resources, and the fact that the assessments are not just about access. Gerwig described them as being about "working with communities to identify health needs, and then using the hospital as a place to have a conversation about how to meet those needs."

In 2016, for the first time, climate and health showed up as one of Kaiser Permanente's prioritized health care needs. Although it did not rank very highly on the list (see Table 6-1), 14 of Kaiser Permanente's 39 hospitals prioritized it.

Even more important, in Gerwig's opinion, is that so many of the other prioritized community health needs have climate co-benefits, thus providing hospitals with a tool to address climate-related health issues by engaging with communities in multiple ways. For example, most facilities listed obesity as the first, second, or third most important community health need (see Table 6-1). Addressing obesity by promoting physical activitiy affects climate action as well—walkable communities, bike-sharing programs, bicycle paths, active transportation, and mass transit all have obesity prevention and climate co-benefits. The same is true of community gardens and reduced meat consumption.

Gerwig next discussed economic security as another example where acting to address a community need could yield a co-benefit. Reiterating that hospitals are big purchasers, she explained that more purchasing from local sources has positive effects both on economic security of a community and the climate. Greater use of local sources increases local jobs, and local jobs mean shorter commutes, which, in turn, reduce fossil fuel use and greenhouse gas emissions. As a final example, she mentioned asthma and remarked that strategies to reduce traffic in order to reduce particulate matter, as a way of addressing asthma, also reduce fossil fuel emissions.

Gerwig used three specific community examples to further illustrate interactions between "climate and health" and the other community health needs. First, she described the rates of youth obesity (percentage obese), ozone (percentage of days exceeding standards), and asthma (percentage of adults with asthma) in Kern County, which is located in California's

TABLE 6-1 2016 Kaiser Permanente Prioritized Community Health Needs

Community Health Need	Number of Kaiser Permanente Hospitals Identifying the Need
Obesity/HEAL[a]/diabetes[b]	39
Behavioral health (mental health and substance abuse/tobacco)	39
Access to care	39
Economic security	31
Violence/injury prevention	30
Cardiovascular disease/stroke	22
Asthma	22
Cancers	21
Oral health	19
HIV/AIDS/sexually transmitted infections	19
Maternal and infant health	15
Climate and health	14
Housing	13
Transportation	11
Education	7

[a] Healthy Eating Active Living.
[b] Community health needs listed in bold are needs that can be addressed through co-benefit strategies (i.e., co-benefits for climate and health).
SOURCE: Gerwig presentation, March 13, 2017.

central valley, as "egregious" compared to either California or national benchmarks. The percentage of youth in Kern County who are obese is 22.41 percent, compared to 18.99 across California; and the percentage of days exceeding ozone standards is 13.54 percent in Kern County, compared to 2.65 percent across California and 1.24 percent nationally. She quoted a community member who participated in the Community Health Needs Assessment: "If you suffer from asthma then you may not go outside and be active and then you are gaining weight and you're not eating healthy food."

As a second example, Gerwig cited youth physical inactivity levels in Moreno Valley, located in southern California. The percentage of Moreno Valley youth who are physically inactive is 45.01 percent, compared to 35.92 percent across California. Again, Gerwig quoted a community member who participated in the assessment: "physical education programs have been scaled back in public schools, and outdoor sports and exercise programs can be challenging because of the hot climate."

Finally, Gerwig cited road density in Riverside County, also in southern California, where the total road network density (road miles per acre) is 5.68, compared to 2.02 across California and 1.45 nationally. Again, she quoted an assessment respondent: "The lack of jobs available in Riverside County also increases commutes for residents, increasing the use of cars on the road and more pollution in the air." For Gerwig, this example illustrates, again, the connection between a lack of local jobs, which means longer commutes, and climate impacts. If there were more local jobs, commutes would shorten, and not only would community economic security improve, but such efforts would have climate co-benefits as well.

Kaiser Permanente's Greenhouse Gas Emissions

Gerwig noted that when discussing Kaiser Permanente's efforts to address climate, she always starts by being transparent about what Kaiser Permanente is emitting. In 2015, the organization's total greenhouse gas emissions amounted to nearly 800,000 metric tons of CO_{2e}.[3] The vast majority of these emissions (91 percent) were from the use of electricity and natural gas to power the buildings and equipment. Specifically, 65 percent of 2015 greenhouse gas emissions was from purchased electricity, 25.9 percent from stationary combustion (natural gas), 0.4 percent from stationary combustion (diesel), 4.3 percent from medical gases, 3.0 percent from refrigerants, and 1.4 percent from fleet vehicles.

Importantly, Gerwig pointed out, Kaiser Permanente's greenhouse gas emissions were down 5 percent in 2015, compared to 2008, despite a 20 percent growth in membership and the construction of more hospitals over this time period.

Having much of their operations in California, Kaiser Permanente draws from a grid that, Gerwig said, "is getting cleaner." Currently, about 25 percent of California's electricity is from renewable power. By 2030, that figure is expected to rise to 50 percent. Because of the political climate in California and the way it allows companies to "green" their energy, Gerwig stated that there are other opportunities in terms of power purchase and other agreements that Kaiser Permanente is able to take advantage of as well.

Kaiser Permanente's Environmental Stewardship Goals: Becoming Carbon Positive by 2025 and Other Climate-Related Goals

Climate action is Kaiser Permanente's number one 2025 environmental stewardship goal. Specifically, the goal is to be net carbon positive by

---

[3] CO_{2e}, or carbon dioxide equivalent, is a standard unit for measuring carbon footprints.

2025. In other words, Gerwig explained, the goal is to become a little bit better than carbon neutral. Several of the other 2025 goals have climate co-benefits. For example, the second 2025 goal is the purchase of 100 percent of Kaiser Permanente's food either locally or from farms and producers that use sustainable practices. Two additional 2025 goals with climate co-benefits are to have zero waste (recycle, reuse, or compost 100 percent of all non-hazardous waste) and to reduce water use (reduce the amount of water used by 25 percent per square foot of buildings).

According to Gerwig, even though Kaiser Permanente emitted nearly 800,000 metric tons of CO_{2e} in 2015, the organization is well on its way to eliminating its greenhouse gas emissions. In 2016, Kaiser Permanente went live with two major off-site energy projects, one solar (110 megawatts [MW] solar power capacity to be purchased from Blythe solar plant in Riverside, California), and the other wind (43 MW wind power capacity to be purchased from turbines at Golden Hills wind farm in Altamont Pass, California). In California, both of these projects are online now and are drastically reducing greenhouse gas emissions. In addition, more than 100 Kaiser Permanente facilities across California are installing onsite solar for a total of 70 megawatts solar power. In sum, Gerwig said, currently half of the electricity that Kaiser Permanente uses to power its California facilities is renewable, consisting of mostly solar or wind sources.

Health Care's Voice on Climate Change

Gerwig closed by adding her thoughts on health care's voice for climate change and why health care institutions are so important in addressing community well-being as it relates to climate. First, she referred to what several other speakers had discussed about how communities are already suffering from the health impacts of climate change and added that, as a health care organization, Kaiser Permanente is seeing this happen. Next, Gerwig referred to earlier discussions on equity and added that Kaiser Permanente appreciates that those who will suffer the most are those who are least able to be resilient against what she described as "the onslaught of climate change." Finally, as big greenhouse gas emitters, with 8 percent of greenhouse gas emissions in the United States coming from health care, Gerwig said, "[w]e have an obligation."

Summary

In summary, Gerwig called for several community health actions around climate change. First, Community Health Needs Assessments provide an important opportunity for communities to engage not just with their health care partners, but with anyone who cares about health, and to address the health impacts of climate change and community vulnerabilities. Second, she encouraged promoting renewable energy and energy efficiency. She recognized that having many of Kaiser Permanente's operations located in California gives her what she described as a "bubble perspective," but she stated there are opportunities in many other parts of the country as well. Third, she encouraged workshop participants to ask their health care partners to publicly report their greenhouse gas emissions. Fourth, she called for Leadership in Energy and Environmental Design (LEED) certification of all hospitals. Finally, she called for suppliers to use local employees.

CALIFORNIA'S INNOVATIVE CLIMATE AND AGRICULTURE POLICIES[4]

The California Climate and Agriculture Network is a coalition of sustainable and organic agriculture organizations that works very closely with farmers and ranchers, scientists who study climate change and agriculture, nonprofit organizations, agricultural professionals, and policy makers, Renata Brillinger began. Brillinger went on to provide an overview of greenhouse gas emissions and other environmental challenges in California agriculture, discuss the impacts of climate change on agriculture, and outline what she described as California's "very ambitious set of policies" related to climate change.

Challenges in California Agriculture: Greenhouse Gas Emissions, Water Contamination, and Other Environmental Problems

About 7 percent of California's greenhouse gas emissions are attributed to on-farm emissions. Twenty-nine percent of these emissions come from enteric fermentation, 28 percent from manure management, 22 percent from soil and crop management, 19 percent from fuel use, and 2 percent from rice cultivation. Brillinger clarified that this 7 percent figure does not include the upstream and downstream impacts of food production, for example, the synthetic fertilizers and other inputs that come onto the farm.

In addition to greenhouse gas emissions, some parts of the state also have a significant groundwater contamination problem from the use of nitrogen as a fertilizer. Both synthetic and organic fertilizers contribute to this problem, Brillinger noted. Parts of the state have significant air quality challenges, notably in the Central Valley. Because of its geography, air polluted by industry, agriculture, and urban development gets trapped.

Without elaborating, Brillinger remarked that there is a "litany of other

---

[4] This section summarizes information presented by Renata Brillinger, co-founder and executive director, California Climate and Agriculture Network, Sacramento, California.

challenges" related to agriculture, including soil salinization, water scarcity, and an alarmingly rapid loss of farmland.

Impacts of Climate Change on Agriculture

Being dependent as it is on natural resources and water, the agriculture sector is arguably on the front lines of the impacts of climate change, Brillinger asserted. Its major impacts on crop production and yields include

- Erratic and extreme weather events, such as the drought that California recently emerged from, followed by possibly the wettest year on record and the anticipation of significant flooding and faster snowpack melting;
- Drought and water scarcity (e.g., 524,000 acres of farmland were fallowed in 2015, causing an estimated $1.84 billion loss);
- New pests and diseases;
- Decreased chill hours (i.e., the number of days below a certain temperature that some fruit and nut trees need in order to properly bear flowers and, subsequently, fruit), which, Brillinger said, is beginning to hit some California crops in a very significant way;
- Subsidence (i.e., the overdrawing of the groundwater table), which led to land sinking up to 2 inches per month in some parts of the Central Valley in 2015, with what Brillinger described as having "astounding" impacts on infrastructure;
- Heat stress for both livestock and farm workers; and
- Economic impacts of all of these factors (e.g., more than 21,000 agriculture jobs were lost in 2015).

California's Climate Strategy

"The good news," Brillinger continued, is that California's governor and legislature have adopted a highly ambitious set of climate policies. The current state mandate is to reduce greenhouse gas emissions to 40 percent below 1990 levels by 2030, a new target that just passed in August 2016 as part of SB (Senate Bill) 32. Among the several specific goals laid out as part of reaching this target, two involve agriculture: carbon sequestration in the land base and reduction of short-lived climate pollutants.

SB 32 was preceded by AB 32, which was passed in 2006 and set an initial target to reduce emissions to 1990 levels by 2020. AB 32 also included a cap-and-trade program, which Brillinger remarked provides economic incentives for many voluntary practices to achieve additional emissions reductions in addition to regulating the largest greenhouse gas-emitting industries. The California program is designed somewhat uniquely, she explained, in that it includes an auction of permits to continue emitting greenhouse gas emissions. The permits are issued at the discretion of the governor and legislature, with money collected from the auction deposited into the Greenhouse Gas Reduction Fund. To date, Brillinger said, approximately $3 billion of the fund has been spent, mostly on activities related to the built environment (i.e., transportation and housing) or forest restoration and protection. Another $2.2 billion has been proposed for the coming fiscal year, 2017–2018. "This is a significant amount of money in a state that was, not so long ago, in debt," Brillinger said.

Using some of the Greenhouse Gas Reduction Fund monies, California has recently established four "climate smart" agricultural programs to give grants to farmers for various activities and projects that reduce greenhouse gases: (1) the Sustainable Agricultural Lands Conservation (SALC) program; (2) the State Water Efficiency and Enhancement Program (SWEEP); (3) the Dairy Methane Reduction Program; and (4) the Healthy Soils Initiative. The funding for these programs has totaled approximately $27 million in 2014–2015, $70 million in 2015–2016, and $85 million (plus an amount for the SALC program, which has yet to be budgeted) in 2016–2017. Brillinger predicted that the total for 2016–2017, when the SALC program is included, will probably be around $110 million.[5] She then went on to describe each of these four programs in detail.

Sustainable Agricultural Lands Conservation

SALC allocates grants for the permanent protection of farmland at risk of development. Brillinger described it as a "sister program" to an affordable housing and smart growth development program and the pairing of the two programs as a "very unique" approach to land planning. Thus far, $42.5 million has been allocated for farmland easements to keep them in production and minimize sprawl development and therefore avoid development-related greenhouse gas emissions. While the budget for the coming year is unclear, again, Brillinger predicts that it will be around $30 million. The "scientific underpinnings" of the program, she said, include evidence showing that one acre of urban land emits 70 times more greenhouse gas than one acre of irrigated cropland.

Brillinger commented that in addition to its impact on climate, the protection of greenbelts around cities also impacts health, for example, by improving air and water quality (i.e., open space and agricultural lands can absorb water and replenish groundwater tables). Additionally, greenbelt protection impacts food security and recreational access to open space.

---

[5] The figure in the budget enacted is $95 million for 2016–2017 (figure provided by Brillinger after the workshop).

State Water Efficiency and Enhancement Program

SWEEP gives direct payments to farmers in the form of competitive grants to reduce water and energy use. This program requires that farmers reduce water use, and because water requires energy to pump, also reduces energy use. Thus far, about $67–$68 million has been allocated to these grants, benefiting more than 500 different farms and projects across the state.

Dairy Methane Reduction

The dairy methane reduction program is complicated and "still taking form," Brillinger said. She described methane as a "very potent greenhouse gas." Thus far, $12 million has been spent on anaerobic digesters on a few of the largest concentrated animal feeding operations in the Central Valley to capture methane and convert it into a bio-gas, and another $50 million has been allocated for this current budget year to be split between anaerobic digesters and alternative manure management practices. CalCAN has been focusing a tremendous amount of energy on the latter, she said, and has been a leading advocate because they would like to see some of the state funding be allocated toward practices that have other benefits in addition to reduced greenhouse gas emissions. They are hoping the funds will be used not only to increase use of alternative manure management practices, specifically composting, but also to move cows onto grass for longer parts of the year.

The confined animal feeding operations in California's Central Valley are "quite harmful" for both the environment and human health, Brillinger said, with big impacts on both air and water quality.

Healthy Soils Initiative

The Healthy Soils Initiative is an even newer program than the dairy methane reduction program. At the time of this workshop, it was still being designed; Brillinger expected it to be rolled out in the next few months. It represents another unique approach to greenhouse gas emissions, in her opinion, because it addresses a powerful benefit of agriculture: that agriculture can be a sink for carbon. She explained that the only places where carbon can go are into the atmosphere or oceans, where, she said, "we don't want it to go," or into forests and agricultural lands, where, she said, "we do want it because it is a building block of life" and "a key component of crop fertility." The goal of the program is to provide grants to farmers to implement practices on their farms that will improve soil health and store more carbon in the soil and in woody plants in the hedgerows around farm boundaries, driveways, and buildings. Brillinger remarked that in addition to storing carbon, these plants also create greater diversity, pollinator habitat, and wildlife corridors for creatures that need to move across farm landscapes. Other activities that this program will likely fund include cover cropping and the use of compost, soil, and mulch, all of which build more carbon into the soil. Building soil carbon, Brillinger said, will reduce dependence on synthetic nitrogen fertilizer, thereby improving both air and water quality, and will result in farm systems that are more resilient to the impact of climate change.

Although the federal Farm Bill touches on some aspects of these four programs, for example, by providing conservation payments to farmers for improved environmental stewardship, these California programs are unique, Brillinger observed. She added that she is aware of farmers who are receiving payments to provide climate benefits by transitioning to climate-smart practices.

In addition to California's climate-smart vision and these four programs in particular, another key piece of the California story, Brillinger said, is the state's attempt to address environmental justice and equity issues. She noted two bills in particular, SB (California Senate Bill) 535 and AB (California Assembly Bill) 197.

SB 535, which passed several years ago, requires that 25 percent of monies spent from the Greenhouse Gas Reduction Fund must be allocated to disadvantaged communities. (Brillinger did not define it, but noted that the state of California defines "disadvantaged" in a specific way.) Brillinger explained that this bill, in part, addresses the disproportionate economic burden that poorer communities face as the state makes its transition to a clean energy economy. It also attempts to address the fact that communities are located in the places where the biggest pollutants are being produced.

AB 197, which passed in August 2016, takes SB 535 one step further, requiring that the state do a better job linking its greenhouse gas reduction efforts with the actual air quality problems created by the same greenhouse gas-emitting industries. "How that actually plays out is an experiment in progress," Brillinger said. "We have quite a bit more work to do."

Even the future of cap-and-trade is in question, according to Brillinger. The program sunsets in 2020. A court case is pending between the state and the oil industry around whether the program is legal, but there are also one or more new bills expected this legislative year that will attempt to expand the program past 2020.

A PERSPECTIVE FROM THE INSTITUTE FOR
TRIBAL ENVIRONMENTAL PROFESSIONALS

Fletcher Wilkinson prefaced his talk by stating that tribal climate change issues are just as diverse as the tribes' locations and cultures around the United States. While he was going to attempt to provide a tribal perspective on climate change, he emphasized that he would not be, by any means, addressing all of the tribes' problems. He then went on to describe ITEP's work in climate change.

ITEP's Climate Change Program

ITEP recently celebrated its 25-year anniversary. The Institute works with tribes on a broad range of environmental issues, from air quality to solid waste. It conducts trainings, provides technical assistance, provides air monitoring equipment, lobbies Congress, and works with partner organizations on projects. Its Climate Change Program was founded in 2009 with an Environmental Protection Agency grant. The intention was to become what Wilkinson called a "one-stop shop" for tribes for anything needed to address climate change. The overarching goal of the program is to help tribes build their capacity to deal with climate change. Wilkinson explained that part of tribal sovereignty is not relying on the federal government, yet addressing climate change issues can be difficult for small tribal communities to do on their own.

One way that ITEP helps build tribal capacity to respond to climate change is through training workshops. ITEP partners with either a host tribe or organization in the region to host the training and then invites 25–35 tribal participants. The workshops are held over the course of a few days, during which workshop participants discuss issues they are facing and how they are addressing those issues. Wilkinson stressed that ITEP works with the tribes through the entire process—from building buy-in and trust from a tribe, to developing a vulnerability assessment, to eventually the tribe writing its own adaptation plan. ITEP has held 30 of these trainings thus far, including one just completed in Anchorage, Alaska; one was scheduled for Spokane later that week, and another four to five were planned for later in the year.

In addition to the trainings, ITEP helps tribes build the capacity to address climate change impacts by maintaining a website with a host of information that tribes can use, such as profiles of other tribes that are working with climate change and examples of tribal adaptation plans that already have been implemented.

ITEP also provides a toolkit that is specific to tribes. Wilkinson described the toolkit as, essentially, a set of documents that tribes can use to walk through the planning process. In addition to other documents, the toolkit includes a template that tribes can use to start an adaptation plan.[6] Along with the toolkit, ITEP provides direct technical assistance to tribes. A few of Wilkinson's colleagues have worked directly with tribes to help them write their adaptation plans.

Tribal Planning Is Different

Wilkinson emphasized that tribal planning is different than a city or state's planning efforts. Many tribal communities, he explained, have been located in one area, even one village, for not hundreds, but thousands of years. The people who live there now grew up there, their grandparents grew up there, their great grandparents grew up there, and so on through many generations. Their entire identity, from their culture to their history, from their spirituality to their food pathways is, Wilkinson said, "all connected to that place where they live." Thus, their sense of place is very strong and, he said, "needs to be integrated throughout the entire planning process."

For example, when a tribe is working on a vulnerability assessment, traditional foods are a big issue that comes up often, Wilkinson observed. Many tribes, especially in the west and in Alaska, do not purchase their foods from supermarkets. Rather, they practice subsistence hunting, gathering, and agriculture. Thus, he said, "traditional foods are really important to them."

In addition to traditional foods, culturally important resources also need to be considered during vulnerability assessments. For example, sagebrush is very important to the Navajo, who use it for ceremonies and a whole host of other purposes. The potential loss of that resource, Wilkinson said, "could be devastating to the culture."

At the adaptation plan stage, again, Wilkinson said, tribal planning is different, with the main difference being tribal reliance on traditional ecological knowledge, or TEK. This knowledge, which comes from having lived in an area for thousands of years, is a powerful tool for tribes that needs to be integrated into their adaptation plans. Wilkinson described TEK as a unique knowledge, or understanding of place, that tribes have accumulated after having observed environmental changes over many generations (e.g., changes in sea ice or wildlife). An example of how this knowledge can be integrated into adaptation plans is predicting how thick sea ice will be in 50–100 years based on knowledge of how thick it was 100 years ago.

Geographic relocation is another factor that is very different for tribes

---

[6] The template is available at http://www7.nau.edu/itep/main/tcc/Resources/adaptation (accessed August 22, 2017).

and something that cities do not face. Wilkinson mentioned a tribe in Louisiana that recently received a grant to move and some tribes in Alaska and Washington that are considering moving. For native people who have lived in an area for so long, with all of their culture based on that location, relocation is, Wilkinson said, "a really emotional thing."

Finally, he noted that coastal erosion is another big climate-related problem that is affecting tribes.

Climate Change and Health

The same connection between climate and health that other speakers have addressed is true of tribes as well, Wilkinson continued. A good example is the food calendar by which many tribes live. That is, tribes collect, harvest, or hunt different types of foods at different times of the year. Plus, some cultural resources have to be collected at specific times. But with climate change, some plant species are moving higher in elevation or disappearing altogether, and the ranges of some animal species are shifting. Wilkinson remarked that he had been in Alaska a few weeks prior to this meeting and was speaking with a man who told Wilkinson that when the man was a child, his tribe would hunt for caribou often. Caribou was still an important food to the man's tribe. But the caribou, the man told Wilkinson, no longer come near the tribe's village. The last time the man went on a caribou hunt, he had to travel 160 miles by snowmobile. "These are some really massive changes that are impacting tribal people," Wilkinson said.

The changes are not just impacting their food. Traveling 160 miles is stressful and taxing on the body. Moreover, not being able to collect this food for one's family and one's community also impacts one's spiritual and cultural connections with the community. Food sharing is an important cultural value. There have been some suicides linked to not being able to provide, according to Wilkinson.

Tribes Are Resilient

Wilkinson concluded with a positive note: tribes are resilient. Because of their long history of adaptation, tribes have several strengths with respect to finding solutions that work for them. Traditional ecological knowledge is one of those strengths, as are tribes' strong cultural and community bonds. He mentioned a program in the Hopi Tribe to teach Hopi farmers new ways to farm their corn, which is becoming a serious challenge on the reservation because of drought. Additionally, he mentioned some communities in Alaska that are holding culture camps for a few weeks in summer to bring elders and youth together to teach language, history, and traditional methods of collecting food. Programs like this, Wilkinson said, strengthen the bonds that small villages rely on for resilience.

DISCUSSION

Following Wilkinson's presentation, he, Gerwig, and Brillinger participated in an open discussion with the audience.

Working with Vulnerable Communities

Moderator Goldman opened the discussion by commenting on the unique relationship with the community that each of the panelists described and how, in each case, particular segments of the community were most impacted by climate change. Specifically, she mentioned Gerwig's observation that people with the least economic opportunity not only drive the furthest to work, but also to purchase things they need because of the lack of supply chains in their communities. Then, in agriculture, as Brillinger discussed, the challenge to change farming practices impacts people inequitably. Finally, with tribes, as Wilkinson discussed, climate change has profound impacts on the traditional gathering of food and on other cultural patterns. Goldman asked the panelists to elaborate on how their respective organizations are working with these more impacted communities and the insights that they have gained from having worked with them.

Gerwig reiterated what she described as the "incredibly useful" platform that Community Health Needs Assessments provide because they draw on listening to the community. Health care providers may think they know what their community's health needs are, but, she said, "we all have our lenses." It is only when they invite conversation and transparency that they see what these needs really are. Additionally, she emphasized the importance of data. For example, the website communitycommons.org/chna (accessed May 9, 2017) has a mapping tool that allows users to enter a zip code and obtain a great deal of information about what is going on in the environment in that area. Having a conversation with a community, coupled with having these data, "will guide the thinking," she said, and ensure a collaborative partnership and holistic consensus.

The core of CalCAN's work, Brillinger said, is guided by their farmer advisors and partners. This is the case for a few reasons. First, farmers are the ones who know what will work. "If it won't work on the ground, if it doesn't keep them in business," she said, no policy, incentive, or amount of funding will "move the needle." Thus, Brillinger and colleagues filter everything they theorize about through their farmer advisors and partners. In her opinion, more needs to be done to ensure that the scientific research being conducted in this area is similarly farmer-led and farmer-participatory. Her guess is that this is true in the public health space as well—that the communities most impacted or most vulnerable are the best people to guide research and planning needs.

Wilkinson concurred that the same is true of tribal programs. He emphasized the importance of "listening to what each tribe needs, because the needs are so different. We let them guide us in how we can help."

When asked how ITEP makes a connection between a tribe and the scientific expertise they might need, for example, to find alternative ways of growing corn, Wilkinson responded that a key part of their training is building partnerships. They like to have a partnership going into each training, and they talk a lot about partnerships during the trainings. For example, for a recent training in Alaska, ITEP partnered with the Alaskan Native Tribal Health Consortium, which Wilkinson described as "true experts" on health issues in Alaska. Having them present at the 3-day training so that everyone could develop rapport with each other was, he said, "really valuable."

Incentives for Innovative Approaches in Agriculture

Jonathan Patz asked Brillinger about incentives for innovative approaches in sustainable agriculture and whether there are special programs for "out-of-the-box" practices. He mentioned entomophagy, the eating of insects (e.g., cricket flour), as a growing area in the field of alternative protein sources that are being studied not just for humans, but also for animal feed (e.g., the feeding of mealworms to chickens). Programs in this area, in Patz's opinion, "could turn into big ticket changes."

Brillinger replied that the programs she discussed are state-run programs and that innovation is constrained by the requirement that every dollar spent be allocated to actually demonstrating reduced greenhouse gas emissions. Even practices less innovative than entomophagy are constrained by this requirement, she noted. This creates a challenge as these innovative programs are designed and rolled out, because there are constraints in what the state is willing and able to fund. Unless it can be shown that a new approach is going to reduce greenhouse gas emissions, she said, "there is nothing" that will fund that approach. The quantitative models being used to project future emissions reductions per practice are based on available science. But given that this is an "emergent" field, she said, the science is limited. Thus, to date, what has been funded is very conservative.

When Patz stressed that an entomophagy approach would, in fact, result in "huge" greenhouse gas reductions, Brillinger suggested that funding such a project would likely have to begin with an animal feed source, then transition to a consumer product.

Kaiser Permanente and the Political Climate in California

George Isham commended Kaiser Permanente's work in the area of climate adaptation as "an inspiration," but described what he called a "conundrum" related to the politicization of the subject of climate change and the overlap between electoral patterns and socioeconomic disparities in states such as California. He wondered what the larger implications of these complex issues are for an organization like Kaiser Permanente when it thinks about how it can contribute to improving community health across California.

Gerwig replied that Kaiser Permanente has adopted a "philosophy of inclusion." That is, as it develops policies, plans, and strategies, it keeps its focus at the level of the community. Their operations in Kern County, California, for example, serve that particular community. The process used to identify that community's health needs is the same that has been used to identify health needs in other communities. "You base it on data," she said. "You base it on what the community members want." The assessments identify specific community issues, such as violence, food scarcity, or something else. In addition to that community-level work, Kaiser Permanente also pursues priorities for the environmental program overall. Gerwig cited renewable energy as an example. The recently built solar plant in Riverside County in southern California, which she had mentioned during her presentation, is not producing electricity only for that community's own use. That electricity, she explained, is going to the grid, thereby supporting reduced fossil fuel use throughout the state.

The Challenges of Communicating About Climate Change with the Agricultural Sector and with Legislature

An unidentified workshop audience member asked Brillinger how CalCAN has communicated with the farming and ranching communities that they work with, which are located primarily in the "red part" of California, and how that communication has changed over time as the agricultural impacts of climate change have become more salient to people in these communities.

Brillinger replied that two changes in particular have advanced CalCAN's work enormously in the past 3 to 4 years. One was the drought, which changed the conversation "quite a bit," she said. "We started talking about drought as much as we talked about climate change." Members of the agricultural community were highly impacted by that and were alarmed, she said. The second was the flow of money from the Greenhouse Gas Reduction Fund into agriculture, beginning in 2014, with the first on-line program being the State Water Efficiency and Enhancement Program.

Brillinger recalled that when CalCAN was founded in 2009, there was no constructive voice of agriculture at the climate policy table in the state of California. At that time, most of the voices were lobbyists opposing AB 32 and trying to limit regulation. Since then, now that nearly $200 million has been allocated, she said, "the tune has changed." She described farmers as "pragmatic." They are not necessarily motivated by greenhouse gas reduction emissions or carbon sequestration, rather by staying in business and making sure they have enough water and other resources. If dollars are available that will get them those benefits, then they will take those dollars. "The demand for these programs is huge," she said, and that demand has helped CalCAN's communication efforts.

Brillinger was also asked by the same audience member how CalCAN has communicated about the connections between climate change and agriculture with people in legislature, who are primarily not from the agricultural sector. Brillinger agreed that the vast majority of legislators in California are urban and that there is a lot of literacy building to do among legislators who do not have a relationship with agriculture. She expressed a desire to work more with the public health sector and others to find multiple mutual benefits related to health, food access, and food security. "We really need to develop some narratives around how this can solve a lot of problems," she said.

Learning from Tribes

Referring to tribes who have lived on their land for thousands of years, passing stories from one generation to the next about how they have adapted to environmental changes in the past, Sanne Magnan asked Wilkinson what could be learned from this oral tradition. Wilkinson replied that while tribes are passing these stories down through oral history, tribes are also protective of these stories. In his opinion, asking what can be learned from these stories about how tribes have adapted in the past is a difficult question. Yet, he said, there are ways that traditional knowledge and science can be brought together for mutual benefit. For example, he mentioned that much of what is known about pre-1950 sea conditions in the Arctic is based on traditional tribal knowledge of sea ice.

An unidentified audience member recalled visiting a Shoalwater Bay Tribe site on the southern peninsula of Washington state, where the tribe has been measuring climate impact for many years. The audience member commented on the impressiveness of the tribe's extensive planning for rising sea water and relocation inland and referred other workshop participants to the Robert Wood Johnson Foundation (RWJF) website, where its story and stories from other communities facing similar experiences are being shared. The tribe was a winner of the RWJF Culture of Health Prize.

Measuring Greenhouse Gas Emissions in Terms of Health

An audience member suggested that given the tremendous harm from carbon offset, particularly in the area of cardiology, perhaps greenhouse gas per unit of delivery should be measured in terms of Healthcare Effectiveness Data and Information Set (HEDIS). Gerwig was unaware of any efforts in that arena, but thought the idea "striking." More broadly, she said, the question explores how more mainstream health care measures can be used in ways that will attract the attention of people working in health care. Goldman added that there is nothing to stop anyone from doing this voluntarily and suggested that a number of organizations could get together and work on it instead of waiting for a regulation.

Reasons for Hope

Goldman closed the discussion by articulating that what she said were two "reasons for hope." First is that when there are situations where people are able to obtain resources, whether through grants or other means, much can be accomplished through community work, even communities not traditionally first in line to address these issues, such as the farming community. Second, in circumstances where it is okay to talk about the reality of climate change and the human role in climate change, again, much can be accomplished. Cricket farming, for example, which Patz had mentioned, may be out-of-the-box today, but, Goldman said, "everything I heard today a decade ago would have been considered to be completely out of the box . . . there was a time when none of this would have been possible to fund with grants or any other way."

7

Reflections on the Day

In the final session of the workshop, Ray Baxter, Frank Loy, and Sanne Magnan reflected on the day's presentations and discussions. Magnan then invited other workshop participants to provide their final comments. This chapter summarizes these concluding thoughts.

REFLECTIONS OF RAY BAXTER[1]

For Ray Baxter, the messages that "rang out" throughout the presentations and panels were that the health effects of climate change are real, they are here now, they are neither fair nor equitable, and they can be prevented.

Regarding the real and present nature of the health effects of climate change, Baxter referred to Jonathan Patz's discussion of the diverse and powerful set of health effects related to urban heat, extreme weather, air pollution, allergens, vector and water-borne disease, the water and food supply effects of climate change, mental health, and climate-related refugees and political instability. Although the ensuing presentations and discussions did not address every instance of any one of these effects, panelists did cover an extraordinarily diverse array of health effects related to climate change and the specific forms that these effects are taking in certain communities. These included heat in Louisville (Panel 1); extreme heat, tick-borne disease, and drinking water in Milwaukee (Panel 2); storms in New York City (NYC) and Baltimore (Panel 3); and deteriorating air quality, drought,

---

[1] This section summarizes closing remarks made by Raymond J. Baxter, former senior vice president, community benefit, research and health policy, Kaiser Permanente.

and groundwater pollution in California and the effects of climate change on traditional foods for native peoples (Panel 4). Additionally, Baxter noted that although fire had not been mentioned, fire and the health effects associated with it have become a regular fact of summer life in the West.

Also with respect to the reality that the health effects of climate change are here now, Baxter referred to Georges Benjamin's "fierce urgency of now" characterization. Baxter said, "We cannot let that fall off the radar screen of the people we are trying to reach."

In addition to the real and extant health effects of climate change, another theme that surfaced throughout the sessions, Baxter continued, is that these effects are experienced inequitably. He recalled that Benjamin, at the beginning of the workshop, pointed out the issue of ethics and morality around inequity and inequitable effects. His remarks were followed by both Patz and the Panel 1 speakers also framing climate change and its health effects as a moral issue. The Milwaukee framework (i.e., Panel 2) also explicitly embraces the issue of equity, Baxter added, as does New Hampshire's use of the Centers for Disease Control and Prevention's Building Resilience Against Climate Effects (BRACE) framework (Panel 3), which identifies vulnerable populations at risk as part of its way of focusing. This is also occurring in Baltimore (Panel 3), where climate resiliency work being done acknowledges how a legacy of racism and segregation still shapes inequity and vulnerability related to climate change.

As a specific example of how climate change impacts populations differently, Baxter mentioned the vulnerability of frail older adults in NYC health care institutions during Hurricane Sandy (Panel 3).

He also called out Patz's description of global inequity—"we are the generators," he said, yet it is the most vulnerable people living in the least developed parts of the world who bear the greatest and first effects of the health effects of climate change.

Regarding the message that the health effects of climate change can be addressed, Baxter referred to Patz's "simple solution," which is to reduce carbon emissions. Doing so immediately creates a healthier environment and can be done, Baxter said, "whether you believe in climate change or not." He added that doing things that address the principal causes of ill health, for example, walking more and eating better, will also help the environment.

The politics are not impossible, Baxter continued. The Kentucky (Panel 1) and California (Panel 4) examples stood out to him as demonstrations that idealistic goals can be reached, and also in the case of Gundersen Health System (Panel 2), that the economic effects can be positive, not just theoretical. Furthermore, the examples in New Hampshire, New York, and Baltimore (Panel 3), among others, demonstrated to Baxter that it is possible to plan and prepare in a comprehensive manner.

Is Local Action Enough?

Two key questions remained for Baxter. The first pertained to local action. Specifically, is local action enough, and can it be scaled? He said that sometimes it seems as though the more locally focused people are, the more optimistic they are. In contrast, the more globally focused they are, the more negative they are.

In his opinion, Kentucky's vision of a just transition to a low-coal or no-coal future could potentially have a big effect by serving as a model for others (Panel 1). "It could demonstrate," he said, "a change that is, in a way, beyond belief." He found it "powerful" that the work being done in Kentucky could have this effect even though their local actions in and of themselves will not change what Baxter called the "global facts" about climate change or its effects on health.

Similarly, Baxter suggested that both the Gundersen and Kaiser Permanente models (Panel 2 and 4) were models that drive the vision for an entire sector of the economy and if scaled, adopted, and spread could make a significant difference well beyond local actions.

Additionally, the comprehensive planning and preparedness examples in New Hampshire, Baltimore, and NYC (Panel 3) were cases that, he said, "defy our common cynicism about our ability to plan together across sectors and address such big problems." He mentioned the NYC example of scaling up from previous hospital cooperation around bioterrorism; in Baltimore, expanding the use of data; and, in New Hampshire, the use of the BRACE framework, which specifically draws on learning from other interventions.

A conclusion Baxter drew from the workshop presentations and discussions was that a "huge gap" remains between local projects and commitments and "the kind of global change that ultimately will be needed." He asked, what happens at the regional level? What are the roles of government and public policy now? How far can unique private projects go? "Climate doesn't respect governmental boundaries," he said, "whether they are local or state or national. So what does that require of us in the approach going forward?"

The Balance Between Prevention and Adaptation

Baxter's second question was: what is the balance between focusing on prevention and mitigation versus focusing on adaptation and resilience? How should this work be prioritized? He found the Louisville example interesting (Panel 1) because of its comprehensive urban heat program that includes not just reforestation, but also addresses carbon reduction and heat generation. He observed that some of this balance seems to depend on whether one is a major generator or extractor. Major generators and extractors potentially have, he said, "very big levers to pull" that can have big effects on prevention. Entities with few levers to pull or with few resources, in contrast, are "almost forced into the resilience and response mode." But again, he asked, is it enough to leave these effects to chance?

Elected officials in particular cannot afford to be second guessed, he continued. NYC has no choice but to prepare for "the next" Irene or Sandy and therefore must focus on resilience (Panel 3). But, he asked, what is its role in generating and contributing to the problem (i.e., the next Irene or Sandy)? He observed that there had been little discussion about the extensive programs aimed at reducing vehicular traffic and increasing mass transit. Yet, in his opinion, those types of programs are probably part of the total and balanced effort that is required.

Again, he cited New Hampshire (Panel 3), this time as an interesting example of a blend of local, topical approaches spread across the state and the use of a comprehensive framework (i.e., the BRACE approach). Baltimore too (Panel 3) has adopted an interesting approach by blending all-hazard mitigation planning with specific climate change planning and using an explicit equity lens, he said, "for all of it." Activities in Milwaukee (Panel 2) and New Hampshire (Panel 3) have elements of this, too. Finally, Kaiser Permanente (Panel 4) uses what Baxter described as a "burden of harm measuring stick" for its prioritization of environmental activities.

Baxter commented on the perception of a shifting locus of responsibility and action for climate (e.g., to local government or to individual communities or private actors). He argued that while it certainly may be the case that local actors are best suited to effective action, especially around resilience, he counted only three workshop speakers who talked about how the emissions and practices of their own institutions contribute to the problem and what their institutions intended to do about it. "We cannot leave ourselves only on one side of the problem," he said. California, he pointed out, is "an outstanding example" of what can be achieved when communities, innovators, businesses, and government policy are aligned around the problem, around the evidence, and around the solutions (Panel 4). But moving away from California and looking broadly across the United States, he asked whether local projects and plans will add up to the regional and global change that is needed to address the fundamental drivers of adverse climate change. Or, he asked, "Will those individual actors and communities and organizations and jurisdictions default almost entirely to disaster preparedness in a narrow, somewhat self-serving kind of resilience planning that is, by force, almost totally defensive and may tend to be largely reserved for the privileged and the advantaged, while the planet at its larger level continues to deteriorate?" The challenge of resilience planning versus mitigation will continue to persist. "Neither one of these is a reasonable choice," he concluded.

REFLECTIONS OF FRANK LOY[2]

Loy agreed with Baxter that although some things were established over the course of the day-long workshop, others were left open. "We have established the seriousness of the problem," Loy began. He referred to Patz's presentation, as well as Benjamin's opening comments, as being particularly helpful in addressing what can be done at both the local and national levels. He questioned, however, why, given what he described as "the overwhelming evidence of the problem and the very thoughtful approaches that we heard," it is so hard to address this problem at the scale it deserves. The answer, in Loy's opinion, is threefold.

First, the problem appears to many people to be in the future. In some places, he said, "you can actually feel it and sense it, in terms of wild fires and water problem[s]." But often, he observed, "you can't, so it looks like something 'out there'."

Second, some people feel that, he said, "it is not really true." This is a serious issue that is exploited by those who benefit from it and who provide, Loy said, "ammunition for that thought." A third factor that "hurts our ability to address this problem," Loy continued, is a sense of hopelessness. The feeling that there is little that can be done about something that is happening now and all over the globe is, he said, a "terrible problem for those of us who are trying to address it."

In addition, Loy stressed the importance of acknowledging that those who care about the problem have made mistakes in the past. For example, as alluded to earlier in the workshop (Panel 1), while, in his opinion, it was right to get rid of coal as a fuel, the problem for those who were going to bear the brunt of the change was not adequately addressed.

Loy also agreed with Lisa Abbott (Panel 1) and other speakers who had stressed the need for process. "You can't just get to where you want to go by being right," he said. "You have to have a process that involves the people." He asserted that experts in the environmental community, and sometimes the health community, are "not as good as we ought to be at really believing in that process and really working that process."

---

[2] This section summarizes closing remarks made by Frank Loy, chair, Roundtable on Environmental Health Sciences, Research, and Medicine and U.S. Representative to the 66th Session of the General Assembly of the United Nations.

Where the Health Community Fits

When people not deeply familiar with the science of climate change want to get a sense of what "really is true and what is baloney," Loy said, they may not be able to seek out scientists for answers, but are likely to turn to trusted leaders from other walks of life who have nothing to do with climate change, he said. These include leaders in the faith community, leaders in the higher education community, and leaders in the health community. Doctors and nurses, Loy said, are respected figures in every community, including communities that have few other respected figures. This is why it is so worthwhile, in his opinion, for the health community to pay attention to this problem. Through their efforts, he said, "one can build the kind of base of support that is needed for action."

Loy mentioned being part of an organization that has analyzed attitudes about climate change among the U.S. population and has found correlations with sex (with women more likely to acknowledge climate change), level of income, and geography. But the most predictive correlation they have found thus far is with party affiliation. The only kind of individual who might be able to impact this correlation, Loy remarked, is not someone who, by reason of their profession, is on one side or the other. Rather, it is someone, he said, who has a "totally different point of entry," such as someone from the health community.

Loy concluded, "From all aspects that I can see, what we are talking about is a public health problem of absolutely first order. This society of public health professionals is the one that can address it most effectively and help us out of this dilemma."

DISCUSSION

Following Loy's remarks, moderator Magnan opened the workshop to anyone who wanted to further reflect on the day's presentations and discussions. George Isham began the discussion by referring to Loy's remarks, as well as other speakers' remarks, around the idea that health will be the "savior" or the winning rationale for addressing environmental challenges. He remarked on the interesting contrast with the way the Roundtable on Population Health Improvement usually thinks about rationales, about ways to identify persuasive rationales to mobilize more effective action in other sectors in response to some of the issues that are being seen in health care and public health.

In addition, Isham reflected on dialogue around climate change. Specifically, in his opinion, dialogue that occurs at the level of party politics typically is not very constructive from either point of view. Yet, when dialogue occurs at a technical, scientific level, it doesn't resonate broadly beyond the technical field of expertise from which it arises. To improve this dialogue, he suggested, first, one should think about the logic model, or framework, in which various climate change factors interact and not be limited by thinking that the outcome is either only health or only the environment. Second, he suggested more communication in "that middle field" between politics and scientific expertise that begins to lay these issues out in broad, more publicly understandable forms that communicate both to the general public and across "technical silos."

Finally, Isham said he had been encouraged and energized by the examples presented at this workshop and hoped that "we have the wisdom" to begin building bridges of communication, not just between the health and environmental sectors, but beyond.

Marthe Gold of The New York Academy of Medicine wondered aloud why climate change has become political and even partisan. She suggested that one reason the messages about the threats of climate change may not be resonating may simply be political affiliation and, she said, "being a good team member." But another reason why the message is not resonating, Gold continued, may be that what is needed is a stronger and more personal way to influence the political discourse. She agreed with Baxter that this is not going to work "simply on a local level."

As far as how to influence politicians and the public, Gold recalled Benjamin's suggestion early in the workshop that the way to motivate people is to provide them with a positive view of what life would be like if the climate were better. In her opinion, that suggestion is optimistic. She explained how prospect theory has taught that, in fact, people respond more strongly to things that they might lose than to things they might gain. Thus, "the sky is falling" may be a better approach. But these different approaches need to be tested, she said. A difficulty with climate is that there is a "great swath of the public" that has not experienced the type of serious climate-related environmental problems, such as flooding or fire, that people living, for example, on the east or west coasts or in low-lying countrysides have experienced. Even though people tend to respond more strongly to things that are going to happen immediately, with climate, they tend to delay. The question then becomes, how can the problem be framed to bring more immediacy? She has always been a proponent of what the media and entertainment industry can contribute and wondered whether there might be approaches in that arena that need to be considered.

Finally, reflecting on Loy's remarks on the role of health professionals, as well as Benjamin's remarks earlier in the workshop, Gold wondered whether there is a way for health care providers to not only inculcate the notion of what people can do individually to promote their health through exercise, diet, and so on, but also talk to their patients about their roles as citizens. She concluded, "We seem to, in this day and age, be less activated as citizens. I wonder what people think and whether the Roundtable might consider whether the health professions have something to say about people being good citizens."

Regarding Loy's remarks on the correlation between attitude toward climate change and party affiliation, Matt Cahillane mentioned a study conducted in New Hampshire showing that, among independents, when asked if they "believed" in anthropogenic climate change, the response depended on the previous day's temperature (Hamilton and Stampone, 2013). On days when the temperature had been similar to the day before, respondents were more likely to respond that they did not believe in human-caused climate change. In contrast, on days when the temperature the day before had been either warmer or cooler, they were, he said, "straight up there with the Democrats in believing in it." For him, these results highlight the short memories and malleability of Americans. He suggested, "maybe we will have the chance with some of those independents."

Gary Gunderson of Stakeholder Health and Wake Forest Baptist Medical Center commented that before this workshop started, he had been anticipating "the most depressing thing ever," but now was "feeling convicted" as a grandfather, as a person of faith, and as a health professional. Although proud of being part of a group of driven, mission-oriented health care systems, he said, "our faith partners have not raised this with any of the clarity that we heard numerous times today." The "prophetic witness" for this, he said, is coming from Kaiser Permanente, which is not a faith-based system, but obviously has a deep moral drive.

Maureen Litchveld offered four key reflections. First, while she considered this to be an enormously successful meeting, she wondered whether it would help the communication process to juxtapose the successes with the failures. While this workshop featured success stories, she suspected that there must be case studies of failures as well. Second, while the local examples presented at this workshop provided good, illustrative ways to characterize assets, she was unsure whether the same characterization of assets has happened at the national level. Third, when considering risk communication and the principles around risk communication, while there was much discussion around who would be good messengers, she stressed the importance of also looking at culture, not just from an ethnic perspective, but also from a community perspective (e.g., community of scientists, community of health professionals, community of policy makers, community of farmers). Working across cultures would be helpful, she remarked. Finally, she offered what she called a "global sense of hope." She mentioned that she would be co-chairing, at end of the April 2017, an expert panel that would be creating a Caribbean-wide roadmap to climate change. Those countries, she said, are taking on the responsibility to act.

In closing, Magnan said that she, too, felt encouraged by the day. When thinking about climate-related disasters and their increasing frequency, she found it compelling to think about an asset-based approach on advancing a health agenda, as opposed to a deficit-based approach. In addition, she felt encouraged by the discussion around communication and the importance of entering conversations not by leading with the issue, in this case, climate change, and perhaps not even with health, but rather with what matters to the constituent. She stressed the importance of asking, "What do you value?" Then, "What do I value? Where is the Venn diagram? Can't we do both?" This communication principle is something, she said, "we are learning over and over again." In her opinion, the adage, "check your agenda at the door," which Halida Hatic had stressed during her presentation (Panel 1), is particularly pertinent to this issue today.

Finally, Magnan echoed the message of hope that Baxter and Loy had expressed. Paraphrasing Jim Collins (Collins, 2001), she said, "Face [the] brutal facts, but never lose hope."

A

References

Battisi, D. S., and R. L. Naylor. 2009. Historical warnings of future food insecurity with unprecedented seasonal heat. *Science* 323(5911):240–244.

Brown, L. 2016, June 28. *Two Baltimores: The White L vs. the Black butterfly*. http://www.citypaper.com/bcpnews-two-baltimores-the-white-l-vs-the-black-butterfly-20160628-htmlstory.html (accessed August 22, 2017).

Caminade, C., J. Turner, S. Metelmann, J. C. Hesson, M. S. Blagrove, T. Solomon, A. P. Morse, and M. Baylis. 2017. Global risk model for vector-borne transmission of Zika virus reveals the role of El Niño 2015. *Proceedings of the National Academy of Sciences of the United States of America* 114(1):119–124.

CDC (Centers for Disease Control and Prevention). 2014. *Assessing health vulnerability to climate change: A guide for health departments*. https://www.cdc.gov/climateandhealth/pubs/assessinghealthvulnerabilitytoclimatechange.pdf (accessed May 9, 2017).

CDC. 2015. CDC's Building Resilience Against Climate Effects (BRACE) Framework. https://www.cdc.gov/climateandhealth/brace.htm (accessed April 28, 2017).

CDC. 2017. *The guide to community preventive services*. https://www.thecommunityguide.org/about/about-community-guide (accessed May 26, 2017).

City of New York. 2013. *A stronger, more resilient New York*. A PlaNYC Report. http://www.nyc.gov/html/sirr/html/report/report.shtml (accessed May 9, 2017).

Collins, J. C. 2001. *Good to great: Why some companies make the leap . . . and others don't.* Random House.

Cook, J., N. Oreskes, P. T. Doran, W. R. L. Anderegg, B. Verheggen, E. W. Maibach, J. S. Carlton, S. Lewandowsky, A. G. Skuce, and S. A. Green. 2016. Consensus on consensus: A synthesis of consensus estimates on human-caused global warming. *Environmental Research Letters* 11(4). http://iopscience.iop.org/article/10.1088/1748-9326/11/4/048002 (accessed May 26, 2017).

Goldman, L., G. Benjamin, S. Hernández, D. Kindig, S. Kumanyika, C. Nevarez, N. R. Shah, and W. Wong. 2016. *Advancing the health of communities and populations: A vital direction for health and health care.* A publication of the National Academy of Medicine's Vital Directions for Health and Health Care policy initiative. https://nam.edu/advancing-the-health-of-communities-and-populations-a-vital-direction-for-health-and-health-care (accessed May 9, 2017).

Gundersen Health System. 2017a. Gundersen health system (GHS) energy efficiency. Figure presented at the workshop Protecting the Health and Well-Being of Communities in a Changing Climate, Washington, DC.

Gundersen Health System. 2017b. Gundersen's road to energy independence: Percent fossil fuel energy use offset. Figure presented at the workshop Protecting the Health and Well-Being of Communities in a Changing Climate, Washington, DC.

Hamer, M., and Y. Chida. 2008. Walking and primary prevention: A meta-analysis of prospective cohort studies. *British Journal of Sports Medicine* 42(4):238–243.

Hamer, M., and Y. Chida. 2009. Physical activity and risk of neurodegenerative disease: A systematic review of prospective evidence. *Psychological Medicine* 39(1):3–11.

Hamilton, L. C., and M. D. Stampone. 2013. Blowin in the wind: Short-term weather and belief in anthropogenic climate change. *Weather, Climate, and Society* 5(2):112–119.

Harris, D. J., G. Atkinson, A. Batterham, K. George, N. T. Cable, T. Reilly, N. Haboubi, and A. G. Renehan. 2009. Lifestyle factors and colorectal cancer risk (2): A systematic review and meta-analysis of associations with leisure-time physical activity. *Colorectal Disease* 11(7):689–701.

Hawken, P., A. Lovins, and L. H. Lovins. 1999, *Natural capitalism: Creating the next industrial revolution.* Snowmass, CO: Rocky Mountain Institute.

IOM (Institute of Medicine). 2015. *Financing population health improvement: Workshop summary.* Washington, DC: The National Academies Press.

IPCC (Intergovernmental Panel on Climate Change). 2013. *Climate Change 2013: The Physical Science Basis. Working Group I Contribution to the Fifth Assessment Report of the Intergovernmental Panel on Climate Change* [T. F. Stocker, D.Qin, G.-K. Plattner, M. Tignor, S. K. Allen, J. Boschung, A. Nauels, Y. Xia, V. Bex and P. M. Midgley (eds.)]. Cambridge University Press, Cambridge, UK and New York, USA. Figure SPM.8(a).

Jeon, C. Y., R. P. Lokken, F. B. Hu, and R. M. van Dam. 2007. Physical activity of moderate intensity and risk of type 2 diabetes: A systematic review. *Diabetes Care* 30(3):744–752.

Kelly, C. P., S. Mohtadi, M. A. Cane, R. Seager, and Y. Kushnir. 2015. Climate change in the Fertile Crescent and implications of the recent Syrian drought. *Proceedings of the National Academy of Sciences of the United States of America* 112(11):3241–3246.

The Lancet. 2015. *Safeguarding human health in the anthropocene epoch: Report of The Rockefeller Foundation–Lancet Commission on planetary health.* http://www.thelancet.com/commissions/planetary-health (accessed June 30, 2017).

Lasinski, M. 2014. *Instagram:@blueagavebalto.* Instagram.com/blueagavebalto (accessed October 9, 2017).

Louisville Center for Health Equity. 2014. *Louisville Metro Health Equity Report 2014: The Social Determinants of Health in Louisville Metro Neighborhoods* (accessed October 9, 2017).

Louisville Office of Sustainability. 2016. *Louisville Urban Heat Management Study.* https://louisvilleky.gov/sites/default/files/sustainability/pdf_files/louisville_heat_mgt_report_final_web.pdf (accessed October 9, 2017).

Louisville Office of Sustainability. 2015. *Louisville Urban Tree Canopy Assessment.* https://louisvilleky.gov/sites/default/files/community_forestry/community_foresty_files/louisville utcreport-24march2015_draft.pdf (accessed October 9, 2017).

Monninkhof, E. M., S. G. Elias, F. A. Viems, I. van der Tweel, A. J. Schuit, D. W. Voskuil, and F. E. van Leeuwen. 2007. Physical activity and breast cancer: A systematic review. *Epidemiology* 18(1):137–157.

NACCHO (National Association of County & City Health Officials). 2008. *Are we ready?: Preparing for the public health challenges of climate change.* http://archived.naccho.org/topics/environmental/climatechange/upload/Are-we-ready_14_view.pdf (accessed May 9, 2017).

NACCHO. 2014. *Are we ready?: Report 2: Preparing for the public health challenges of climate change.* https://www.naccho.org/uploads/downloadable-resources/NA609PDF-AreWeReady2.pdf (accessed May 9, 2017).

NASEM (National Academies of Sciences, Engineering, and Medicine). 2016. *Attribution of extreme weather events in the context of climate change.* Washington, DC: The National Academies Press.

NCD-RIsC (NCD Risk Factor Collaboration). 2016. Trends in adult body-mass index in 200 countries from 1975 to 2014: A pooled analysis of 1698 population-based measurement studies with 19.2 million participants. *The Lancet* 387(10026):1377–1396.

New Hampshire Department of Health and Human Services. 2017. 2011–2015 social vulnerability overview. New Hampshire Public Health Data.

NRC (National Research Council). 2011. *America's climate choices.* Washington, DC: The National Academies Press.

NRC. 2012. *Climate change: Evidence, impacts, and choices: PDF booklet.* Washington, DC: The National Academies Press.

NRC. 2013. *Abrupt impacts of climate change: Anticipating surprises.* Washington, DC: The National Academies Press.

Patz, J. A., M. A. McGeehin, S. M. Bernard, K. L. Ebi, P. R. Epstein, A. Grambsch, D. J. Gubler, P. Reither, I. Romieu, J. B. Rose, J. M. Samet, and J. Trtanj. 2000. *The potential health impacts of climate variability and change for the United States: Executive summary of the report of the health sector of the US National Assessment. Environmental Health Perspectives* 108(4):367.

Patz, J., H. K. Gibbs, J. A. Foley, J. V. Rogers, and K. R. Smith. 2007. Climate change and global health: Quantifying a growing ethical crisis. *EcoHealth* 4(4):397–405.

Patz, J. A., S. J. Vavrus, C. K. Uejio, and S. L. McLellan. 2008. Climate change and waterborne disease risk in the Great Lakes region of the U.S. *American Journal of Preventive Medicine* 35(5):451–458.

Patz, J. A., H. Frumkin, T. Holloway, D. J. Vimont, and A. Haines. 2014. Climate change: Challenges and opportunities for global health. *Journal of the American Medical Association* 312(15):1565–1580.

Pucher, J., R. Buehler, D. R. Bassett, and A. L. Dannenberg. 2010. Walking and cycling to health: A comparative analysis of city, state, and international data. *American Journal of Public Health* 100(1):1986–1992.

Reuters. 2015. Maps income per capita, unemployment and percent of African Americans by block groups in Baltimore. 30 cm wide (sin01), edited by RTX1AR1D.eps.

Scarborough, P., P. N. Appleby, A. Mizdrak, A. D. M. Briggs, R. C. Travis, K. E. Bradbury, and T. J. Key. 2014. Dietary greenhouse gas emissions of meat-eaters, fish-eaters, vegetarians and vegans in the UK. *Climatic Change* 125(2):179–192.

Thompson, J. 2017. *Lead true.* Charleston, SC: ForbesBooks.

Thompson, T. M., S. Rausch, R. K. Saari, and N. E. Selin. 2014. A systems approach to evaluating the air quality co-benefits of U.S. carbon policies. *Nature Climate Change* 4:917–923.

Toner, E. S., M. McGinty, M. Schoch-Spana, D. A. Rose, M. Watson, E. Echols, and E. G. Carbone. 2017. A community checklist for health sector resilience informed by Hurricane Sandy. *Health Security* 15(1):53–69.

Van Panhuis, W. G., M. Choisy, X. Xiong, N. S. Chok, P. Akarasewi, S. Iamsirithaworn, S. K. Lam, C. K. Chong, F. C. Lam, B. Phommasak, and P. Vongphrachanh. 2015. Region-wide synchrony and traveling waves of dengue across eight countries in Southeast Asia. *Proceedings of the National Academy of Sciences of the United States of America.* 112(42):13069–13074.

West, J.J ., S. J. Smith, R. A. Silva, V. Naik, Y. Zhang, Z. Adelman, M. M. Fry, S. Anenberg, L. W. Horowitz, J. F. Lamarque. 2013. Co-benefits of Global Greenhouse Gas Mitigation for Future Air Quality and Human Health. *National Climate Chang* 3(10):885–889.

Westhoek, H., J. P. Lesschen, T. Rood, S. Wagner, A. De Marco, D. Murphy-Bokern, A. Leip, H. van Grinsven, M. A. Sutton, O. Oenema. 2014. Food choices, health and environment: Effects of cutting Europe's meat and dairy intake. *Global Environmental Change* 26:196–205.

Wisconsin Climate and Health Program. 2016. *Climate and health community engagement toolkit: Planning guide for public health and emergency response professionals.* https://www.dhs.wisconsin.gov/publications/p01637.pdf (accessed May 9, 2017).

Woodcock, J., P. Edwards, C. Tonne, B. G. Armstrong, O. Ashiru, D. Banister, S. Beevers, Z. Chalabi, Z. Chowdhury, A. Cohen, O. H. Franco, A. Haines, R. Hickman, G. Lindsay, I. Mittal, D. Mohan, G. Tiwari, A. Woodward, and I. Roberts. 2009. Public health benefits of strategies to reduce greenhouse-gas emissions: urban land transport. *The Lancet* 374(9705):1930–1943.

Ziska, L., K. Knowlton, C. Rogers, National Allergy Bureau, Aerobiology Research Laboratories, Canada. 2016 update to data originally published in L. Ziska, K. Knowlton, C. Rogers, D. Dalan, N. Tierney, M. Elder, W. Filley, J. Shropshire, L. B. Ford, C. Hedberg, P. Fleetwood, K. T. Hovanky, T. Kavanaugh, G. Fulford, R. F. Vrtis, J. A. Patz, J. Portnoy, F. Coates, L. Bielory, and D. Frenz. 2011. Recent warming by latitude associated with increased length of ragweed pollen season in central North America. *Proceedings of the National Academy of Sciences of the United States of America* 108:4248–4251.

ADDITIONAL REFERENCES AND RESOURCES

The following references are cited in materials provided to workshop participants, but not directly cited during the workshop.

Additional References and Resources from the Communities and Organizations Represented at This Workshop

Bliss, L. 2017. Louisville's faith-based plan to fight urban heat. *The Atlantic, City Lab.* http://www.citylab.com/cityfixer/2017/02/louisville-is-beating-the-heat-with-spirituality/515880 (accessed May 9, 2017).

CDC (Centers for Disease Control and Prevention). 2014. *Climate models and the use of climate projections: An overview for health departments.* https://www.cdc.gov/climateandhealth/pubs/climate_models_and_use_of_climate_projections.pdf (accessed May 9, 2017).

CDC. 2015. *Building Resilience Against Climate Effects (BRACE) framework.* https://www.cdc.gov/climateandhealth/BRACE.htm (accessed May 9, 2017).

The Climate Reality Project. 2017. Climate and Health Meeting. Archived webcast. Atlanta, GA. https://www.climaterealityproject.org/health (accessed May 9, 2017).

Crimmins, A., J. Balbus, J. Gamble, C. Beard, J. Bell, D. Dodgen, R. Eisen, N. Fann, M. Hawkins, and S. Herring. 2016. The impacts of climate change on human health in the United States: A scientific assessment. *Global Change Research Program: Washington, DC, USA.* https://health2016.globalchange.gov (accessed August 21, 2017).

Georgia Institute of Technology Urban Climate Lab. 2016. *Louisville Urban Heat Management Project.* Draft for Public Comment. https://louisvilleky.gov/sites/default/files/advanced_planning/louisville_heat_mgt_revision_final_prelim.pdf (accessed May 9, 2017).

Gould, S., and L. Rudolph. 2014. *Why we need climate, health, and equity in all policies.* National Academy of Medicine Commentary. https://nam.edu/perspectives-2014-why-we-need-climate-health-and-equity-in-all-policies (accessed May 9, 2017).

New Hampshire Department of Environmental Services, Climate Change Policy Task Force. 2009. *The New Hampshire Climate Action Plan.* https://www.des.nh.gov/organization/divisions/air/tsb/tps/climate/action_plan (accessed May 9, 2017).

PHI (Public Health Institute). 2017. *A physician's guide to climate change, health and equity.* http://climatehealthconnect.org/resources/physicians-guide-climate-change-health-equity/ (accessed May 9, 2017).

Additional Relevant Reports, Journal Articles, and Other Documents

Abt Associates. 2016. *Climate adaptation the state of practice in U.S. communities.* A report for the Kresge Foundation. http://kresge.org/sites/default/files/library/climate-adaptation-the-state-of-practice-in-us-communities-full-report.pdf (accessed May 9, 2013).

Buonocore, J. J., P. Luckow, G. Norris, J. D. Spengler, B. Biewald, J. Fisher, and J. I. Levy. 2016. Health and climate benefits of different energy efficiency and renewable energy choices. *Nature Climate Change* 6:100–105.

Health & Environmental Funders Network. 2015. *Achieving a climate for health: Philanthropy to promote health and justice through the challenges of climate change.* http://www.hefn.org/sites/default/files/uploaded_files/report_health_and_climate_philanthropy_hefn_ecoamerica_web.pdf (accessed May 9, 2017).

Kennedy, C. A., I. Stewart, A. Facchini, I. Cersosimo, R. Mele, B. Chen, M. Uda, A. Kansal, A. Chiu, K. Kim, C. Dubeux, E. Lebre La Rovere, B. Cunha, S. Pincetl, J. Keirstead, S. Barles, S. Pusaka, J. Gunawan, M. Adegbile, M. Nazariha, S. Hoque, P. J. Marcotullio, F. G. Otharán, T. Genena, N. Ibrahim, R. Farooqui, G. Cervantes, and A. D. Sahin. 2015. Energy and material flows of megacities. *Proceedings of the National Academy of Sciences of the United States of America* 112(19):5985–5990.

Koopman, M. E., and T. Graham. 2015. Whole community adaptation to climate change. *Reference Module in Earth Systems and Environmental Sciences.* https://doi.org/10.1016/B978-0-12-409548-9.09366-0.

National Academy of Sciences and The Royal Society of London. 2014. *Climate change: Evidence and causes.* http://nas-sites.org/americasclimatechoices/events/a-discussion-on-climate-change-evidence-and-causes (accessed May 9, 2017).

Rosenberg, J. 2012. U.S. Navy bracing for climate change. *Global climate change: Vital signs of the planet* (a National Aeronautics and Space Administration publication). https://climate.nasa.gov/news/699/us-navy-bracing-for-climate-change (accessed May 9, 2017).

Stoett, P., P. Daszak, C. Romanelli, C. Machalaba, R. Behringer, F. Chalk, S. Cornish, S. Dalby, B. F. de Souza Dias, and Z. Iqbal. 2016. Avoiding catastrophes: Seeking synergies among the public health, environmental protection, and human security sectors. *The Lancet Global Health* 4(10):e680–e681.

Tabuchi, H. 2017. In America's heartland, discussing climate change without saying "climate change." *The New York Times.* January 28. https://www.nytimes.com/2017/01/28/business/energy-environment/navigating-climate-change-in-americas-heartland.html?_r=0 (accessed May 9, 2017).

Universal Ecological Fund. 2016. *The truth about climate change.* https://feu-us.org/the-report/ (accessed May 9, 2017).

U.S. Department of Agriculture. 2017. *National drought resilience partnership: 2016 end of year report.* https://www.usda.gov/sites/default/files/documents/ndrp-january-2017-end-of-year- report.pdf (accessed May 9, 2017).

U.S. Navy. 2010. *U.S. Navy climate change roadmap.* http://www.navy.mil/navydata/documents/CCR.pdf (accessed May 9, 2017).

WAMU Radio. 2017. Conservatives make the case for action on climate change. Archived audio, plus context and commentary from Twitter. http://the1a.org/shows/2017-02-16/conservatives-make-the-case-for-action-on-climate-change (accessed May 9, 2017).

B

Workshop Agenda

**Protecting the Health and Well-Being of
Communities in a Changing Climate
March 13, 2017**

**National Academy of Sciences Building, Auditorium
2101 Constitution Avenue, NW, Washington, DC**

WORKSHOP OBJECTIVES

In the context of considerations of health equity, economic viability, social acceptability, political palatability, and regional fit, workshop participants will:

1. Receive an overview of the health implications of climate change;
2. Explore mitigation/prevention and adaptation/resilience-building strategies deployed by different sectors at various levels (e.g., local, national) and in various regions; and
3. Discuss aspects of collaboration on climate and health issues among community-based organizations, health care systems, businesses, and public health and other local government agencies, along with lessons learned.

8:30 am   Welcome and Opening Remarks
               Lynn Goldman, Dean, Milken Institute School of
               Public Health, The George Washington University

               George Isham, Senior Advisor, HealthPartners; Senior
               Fellow, HealthPartners Institute for Education and
               Research; Co-Chair, Roundtable on Population Health
               Improvement

               Georges Benjamin, Executive Director, American
               Public Health Association

9:00 am   **Setting the Stage**
               Jonathan Patz, Director, Global Health Institute;
               Nelson Institute, Center for Sustainability and the
               Global Environment (SAGE); Population Health
               Sciences; University of Wisconsin–Madison

9:30 am   **Q&A and Discussion**
               Moderator: Henry (Andy) Anderson, Adjunct
               Professor, Population Health, University of
               Wisconsin–Madison; Former Chief Medical Officer,
               Wisconsin Division of Public Health; Planning
               Committee Member

10:00 am   **NETWORKING BREAK**

10:15 am   **Panel 1: Regional Perspectives from the South**
               Moderator: Linda Rudolph, Director, Climate Change
               and Public Health Project, Public Health Institute;
               Planning Committee Member

               Halida Hatic, Director, Community Relations and
               Development, Center for Interfaith Relations,
               Louisville, Kentucky; and Rachel Holmes, Healthy
               Trees, Healthy Cities Program Coordinator, The
               Nature Conservancy

               Maria Koetter, Director, Office of Sustainability,
               Louisville Metro Government

               Lisa Abbott, Empower Kentucky Organizer,
               Kentuckians for the Commonwealth

11:05 am   **Q&A and Discussion**
Moderator: Linda Rudolph

11:35 am   **Panel 2: Regional Perspectives from the Midwest**
Moderator: Surili Patel, Senior Program Manager, Environmental Health, American Public Health Association

Paul A. Biedrzycki, Director, Disease Control and Environmental Health, City of Milwaukee Health Department

Jeff Thompson, Executive Advisor and Chief Executive Officer Emeritus, Gundersen Health System, Wisconsin

12:10 pm   **Q&A and Discussion**
Moderator: Surili Patel

12:40 pm   **LUNCH** (provided for members and speakers; National Academy of Sciences cafeteria and neighborhood options available to attendees)

1:40 pm   **Panel 3: Regional Perspectives from the Northeast**
Moderator: Paul A. Biedrzycki, Director, Disease Control and Environmental Health, City of Milwaukee Health Department

Celia Quinn, Director, Bureau of Health Care System Readiness, New York City Department of Health and Mental Hygiene

Matt Cahillane, Program Manager, Bureau of Public Health Protection, Division of Public Health, New Hampshire Department of Health and Human Services

Kristin Baja, Climate and Resilience Planner, Office of Sustainability, Baltimore City

2:30 pm   **Q&A and Discussion**
Moderator: Paul A. Biedrzycki

3:00 pm   **BREAK**

3:15 pm   **Panel 4: Regional Perspectives from the West**
               Moderator: John Bolduc, Environmental Planner,
               Cambridge (MA) Community Development
               Department

               Kathy Gerwig, Vice President, Employee Safety,
               Health and Wellness; Environmental Stewardship
               Officer, Kaiser Permanente

               Renata Brillinger, Executive Director, California
               Climate and Agriculture Network (CalCAN)

               Fletcher Wilkinson, Climate Change Program
               Coordinator, Institute for Tribal Environmental
               Professionals, Northern Arizona University

4:05 pm   **Q&A and Discussion**
               Moderator: John Bolduc

4:30 pm   **Closing Remarks and Reflections on the Day**
               Ray Baxter, Health Policy Advisor
               Sanne Magnan, Co-Chair, Roundtable on Population
               Health Improvement
               Frank Loy, Chair, Roundtable on Environmental
               Health Sciences, Research, and Medicine

5:15 pm   **ADJOURN**

C

Biosketches of Presenters
and Moderators

Lisa Abbott, M.P.P., is the Empower Kentucky Organizer at Kentuckians for the Commonwealth (KFTC), where she coordinates the organization's effort to develop a "people's energy plan" for Kentucky. She has worked as a community organizer with KFTC for more than two decades, including 7 years spent organizing in coalfields of eastern Kentucky and 14 years as KFTC's organizing director. She serves on the board of the New World Foundation. Ms. Abbott holds a B.S. in biology from the University of North Carolina at Chapel Hill. She received a master's degree in public policy with an emphasis in Leadership Development from the University of Maryland.

Henry A. Anderson, M.D., is an adjunct professor of population health at the University of Wisconsin Medical School. He was previously the state health officer, chief medical officer, and state epidemiologist for occupational and environmental health in the Wisconsin Division of Public Health. His expertise includes public health; preventive, environmental, and occupational medicine; respiratory diseases; epidemiology; human health risk assessment; and risk communication. His research interests include disease surveillance, risk assessment, health hazards of Great Lakes sport-fish consumption, arsenic in drinking water, asbestos disease, and occupational fatalities and injuries. He is certified by the American Board of Preventive Medicine with a subspecialty in Occupational and Environmental Medicine and is a fellow of the American College of Epidemiology. Dr. Anderson is past chair of the Board of Scientific Councilors of the National Institute for Occupational Safety and Health. He has served on several National

Research Council (NRC) committees, including the Division on Earth and Life Studies Committee, the Committee on Toxicity Testing and Assessment of Environmental Agents, and the Committee on Enhancing Environmental Health Content in Nursing Practice. Dr. Anderson received his M.D. from the University of Wisconsin Medical School.

Kristin Baja, M.S., M.U.P., is the climate and resilience planner for the Office of Sustainability of the City of Baltimore. She is responsible for the development and implementation of the city's Disaster Preparedness Project and Plan (DP3), which integrates climate adaptation with hazard mitigation efforts. She is also responsible for climate change communication and outreach, Community Rating System certification, resiliency planning, and STAR Communities certification. Ms. Baja, a certified floodplain manager, regulates the city's floodplain. She is an active member of the Urban Sustainability Directors Network, Climate Communications Consortium of Maryland, American Society of Adaptation Professionals, and the Baltimore City Forestry Board. Previously, Ms. Baja developed the City of Ann Arbor's Climate Action Plan and Sustainability Framework. She has been involved in climate and resilience planning with various cities throughout the United States. In 2016, Ms. Baja was recognized by the White House as a Champion of Change for her work on climate equity and resilience. Ms. Baja holds a master's of urban planning degree and a master of science in Natural Resources and Environment: Sustainable Systems degree from the University of Michigan.

Raymond Baxter, Ph.D., was Kaiser Permanente's senior vice president for Community Benefit, Research and Health Policy for the last 15 years. In that role, he built the largest U.S. community benefit program, investing more than $2 billion annually in community health. He led Kaiser Permanente's signature national health improvement partnerships, including the Weight of the Nation, EveryBody Walk!, the Convergence Partnership, and the Partnership for a Healthier America. He established Kaiser Permanente's Center for Effectiveness and Safety Research and its national genomics research bank, served as President of KP International and chaired Kaiser Permanente's field-leading environmental stewardship work.

Previously, Dr. Baxter headed the San Francisco Department of Public Health, the New York City Health and Hospitals Corporation, and The Lewin Group. In 2001 the University of California (UC), Berkeley, School of Public Health honored him as a Public Health Hero for his service in the AIDS epidemic in San Francisco. In 2006 he received the Centers for Disease Control and Prevention (CDC) Foundation Hero Award for addressing the health consequences of Hurricane Katrina in the Gulf Coast. He currently serves on the advisory board of the UC Berkeley School of

Public Health and the boards of the CDC Foundation and the Blue Shield of California Foundation. He recently completed terms on the Global Agenda Council on Health of the World Economic Forum, and the National Academies of Sciences, Engineering, and Medicine's Roundtable on Population Health Improvement. He holds a doctorate from Princeton University.

Georges Benjamin, M.D., is known as one of the nation's most influential physician leaders because he speaks passionately and eloquently about the health issues that have the most impact on our nation today. From his first-hand experience as a physician, he knows what happens when preventive care is not available and when the healthy choice is not the easy choice. As executive director of the American Public Health Association (APHA) since 2002, he is leading APHA's push to make the United States the healthiest nation in one generation.

He came to APHA from the Maryland Department of Health and Mental Hygiene. Dr. Benjamin became Secretary of Health in Maryland in 1999, following 4 years as Deputy Secretary for Public Health Services. As Secretary, Dr. Benjamin oversaw the expansion and improvement of the state's Medicaid program.

Dr. Benjamin is a graduate of the Illinois Institute of Technology and the University of Illinois College of Medicine. He is board certified in internal medicine and a fellow of the American College of Physicians, a fellow of the National Academy of Public Administration, a fellow emeritus of the American College of Emergency Physicians, and an honorary fellow of the Royal Society of Public Health in Great Britain.

Paul Biedrzycki, M.P.H., M.B.A., CIH, serves as the director of disease control and environmental health within the City of Milwaukee Health Department (MHD). He has worked in local public health for more than 33 years and currently provides leadership and strategic oversight across a wide range of programs and initiatives at MHD. He has been active in the response to a number of emerging infectious disease outbreaks in the City of Milwaukee and surrounding regions that were national in scope and impact. These include the 1993 *Cryptosporidium* outbreak, the 2003 Monkeypox outbreak, the 2006 multistate *E. coli* spinach outbreak, a 2008 nationwide measles outbreak, and the 2009 H1N1 influenza pandemic.

Mr. Biedrzycki has presented extensively at national conferences, workshops, and seminars on a variety of emerging public health topics. These include global climate change, biosurveillance and intelligence fusion, novel infectious disease epidemiology, legal preparedness, and community engagement in emergency planning and response. Mr. Biedrzycki routinely participates in several national workgroups and committees convened by federal agencies such as the Department of Homeland Security, Centers for Disease Control and Prevention, and Environmental Protection Agency.

John Bolduc, MUP, has been an environmental planner for the City of Cambridge, Massachusetts, for 20 years and has more than 30 years of experience in municipal sustainability policy. His professional experience includes local climate change planning for mitigation and adaptation, energy efficiency building, renewable energy, green buildings, open-space planning and management, wetlands protection, and land use policy. Currently, he is managing the Cambridge Climate Change Vulnerability Assessment and Climate Change Preparedness and Resilience Plan. Mr. Bolduc holds a B.S. from the University of California, Davis, and a master's in Urban and Environmental Policy from Tufts University.

Renata Brillinger is the co-founder and executive director of the California Climate and Agriculture Network (CalCAN). She has 18 years of experience in sustainable agriculture policy and food systems projects and has held numerous nonprofit administrative positions since 1992. Prior to CalCAN, she was the program director at the Climate Protection Campaign, focused on agriculture and renewable energy. For 7 years she served as the director of Californians for GE-Free Agriculture, a coalition of sustainable agriculture and environmental organizations that provided education on genetic engineering in agriculture. She serves on the steering committee of the Center for Sustainability at California Polytechnic State University in San Luis Obispo and the advisory board of the University of California, Davis, Agricultural Sustainability Institute.

Matt Cahillane is the public health program manager at the Bureau of Public Health Protection of the New Hampshire Department of Health and Human Services. Mr. Cahillane has managed a number of state-level environmental health programs in New Hampshire over the past two decades. His current projects include building resilience against the effects of a changing climate and local-level interventions for heat stress among the elderly and tick exposure among youth. His past projects have included hazardous materials response, indoor air quality, and disease surveillance.

Kathy Gerwig, M.B.A., is the vice president of employee safety, health, and wellness, as well as environmental stewardship officer, at Kaiser Permanente. She is responsible for developing, organizing, and managing a nationwide environmental initiative for Kaiser Permanente. Her book, *Greening Health Care: How Hospitals Can Heal the Planet,* examines the intersections of health care, health, and environmental stewardship. Ms. Gerwig is also responsible for eliminating workplace injuries, promoting healthy lifestyle choices, and reducing health risks for the organization's 220,000 employees and physicians.

Lynn Goldman, M.D., M.P.H., is the Michael and Lori Milken Dean at the Milken Institute School of Public Health at The George Washington (GW) University. Dr. Goldman's responsibilities are informed by her broad and deep public policy and academic experience. Prior to joining GW in 2010, she was professor of environmental health sciences at the Johns Hopkins Bloomberg School of Public Health. Dr. Goldman was assistant administrator for toxic substances at the Environmental Protection Agency (EPA) from 1993 through 1998 under President Bill Clinton. Under her watch, EPA overhauled the nation's pesticide laws, expanded right-to-know requirements for toxin release, reached consensus on an approach to testing chemicals with endocrine-disrupting potential, developed standards to implement lead-screening legislation, and promoted children's health and global chemical safety. Prior to joining EPA, Dr. Goldman worked in environmental health for the California Department of Public Health.

A member of the National Academy of Medicine, she has chaired or served on numerous committees and forums. She currently serves on the National Academy of Medicine Governing Council and the Governing Board of the National Academy of Sciences. She serves as a member of the Advisory Committee to the Director of the Centers for Disease Control and Prevention and a member of the Food and Drug Administration Science Board.

Among many accolades, Dr. Goldman received a 2009 Heinz Award, given to innovators addressing global change caused by the impact of human activities. She was awarded alumna of the year by the University of California, Berkeley, School of Public Health; received the Woodrow Wilson Award for Excellence in Government from Johns Hopkins University; and was named 1 of 150 outstanding alumni by the University of California, San Francisco. She also received an honorary doctorate from Örebro University in Sweden for her contributions to chemical legislation in the United States and Sweden and her influence on the research conducted at the university's Man Technology Environment Research Centre.

Halida Hatic, M.A., is the director of community relations and development for the Center for Interfaith Relations (CIR), a Louisville-based nonprofit working to promote and support interfaith understanding, cooperation, and action. Before joining the CIR staff, Ms. Hatic worked for 10 years in the environmental sustainability field, including 6 years at the New England office of the Environmental Protection Agency (EPA). In this capacity, she managed and coordinated regional implementation of EPA's voluntary programs to reduce emissions from mobile sources.

In her current role as director of community relations and development for the Center for Interfaith Relations, Ms. Hatic leads efforts to infuse principles of environmental health and sustainability into CIR's daily operations and ongoing programming. She is a certified Green Specialist, a certified consultant in the Barrett Cultural Transformation Tools, Levels I and II, and a 2014 graduate of the Bioneer's Cultivating Women's Leadership program. Ms. Hatic currently serves as vice chair of the board of the Louisville Sustainability Council and has infused her passion and commitment to protecting the health and well-being of all living things into both her personal and professional life. Ms. Hatic holds a B.A. in International and Development Economics from the University of New Hampshire and a master's in Urban and Environmental Policy and Planning from Tufts University.

Rachel Holmes, M.F., is the coordinator of The Nature Conservancy's national urban forestry program called the Healthy Trees, Healthy Cities Initiative. In this role, she educates and mobilizes people nationwide to engage in long-term urban forest stewardship. She also currently leads a conservancy-wide initiative to enhance and expand the organization's efforts to engage youth and communities in urban conservation worldwide, with a special focus on human health and well-being. Particularly passionate about building strategic partnerships with communities of faith, Ms. Holmes is co-developing an urban conservation engagement/ecological data collection protocol with the Center for Interfaith Relations (in Louisville, Kentucky) called the Landscape Audit for Sacred Spaces. Prior to joining the Conservancy, Ms. Holmes created and implemented two urban forestry-based job training programs for urban youth in Connecticut and served as an urban forester for the state of Connecticut's Department of Energy and Environmental Protection. She holds a B.S. of Science from Rutgers University, and concurrently earned a master's of Divinity from the Yale Divinity School and a master's of Forestry from the Yale School of Forestry and Environmental Studies.

George Isham, M.D., M.S., is the senior advisor to HealthPartners, responsible for working with the board of directors and the senior management team on health and quality-of-care improvement for patients, members, and the community. Dr. Isham is also a Senior Fellow at HealthPartners Research Foundation and facilitates forward progress at the intersection of population health research and public policy. He is nationally active and currently co-chairs the National Quality Forum–convened Measurement Application Partnership, chairs the National Committee for Quality Assurance's (NCQA's) clinical program committee, and is a member of NCQA's Committee on Performance Measurement. He is a former member of the

Centers for Disease Control and Prevention's (CDC's) Task Force on Community Preventive Services and the Agency for Healthcare Research and Quality's U.S. Preventive Services Task Force. He currently serves on the advisory committee to the director of CDC. His experience as a general internist was with the Navy, at the Freeport Clinic in Freeport, Illinois, and as a clinical assistant professor of medicine at the University of Wisconsin Hospitals and Clinics in Madison, Wisconsin. In 2014 Dr. Isham was elected to what is now the National Academy of Medicine. He has chaired three studies in the National Academies of Sciences, Engineering, and Medicine's Health and Medicine Division (HMD, the program unit of the former Institute of Medicine) in addition to serving on a number of HMD studies related to health and quality of care. In 2003 Dr. Isham was appointed as a lifetime National Associate of the National Academies in recognition of his contributions to the work of HMD.

Maria Koetter is Louisville, Kentucky's first director of sustainability and is responsible for city-wide strategic sustainability planning, policy development, and program implementation. Ms. Koetter developed Louisville's comprehensive sustainability plan, "Sustain Louisville," which was released in 2013. Key programs under her direction focus on energy conservation, green infrastructure, tree canopy, and urban heat island mitigation. Ms. Koetter formerly worked in the environmental and sustainability consulting industry and has extensive experience with corporate social responsibility and organizational sustainability funding.

Sanne Magnan, M.D., Ph.D., is the co-chair of the Roundtable on Population Health Improvement. Dr. Magnan served as president and CEO of the Institute for Clinical Systems Improvement (ICSI) until January 2016. She was previously the president of ICSI when she was appointed by former Minnesota Governor Tim Pawlenty to serve as commissioner of health for the Minnesota Department of Health. She served in that position from 2007 to 2010 and had significant responsibility for the implementation of Minnesota's 2008 health reform legislation, including the Statewide Health Improvement Program (SHIP), standardized quality reporting, development of provider peer grouping, certification process for health care homes, and baskets of care. She returned as ICSI's president and CEO in 2011. Dr. Magnan also currently serves as a staff physician at the Tuberculosis Clinic at the St. Paul-Ramsey County Department of Public Health and as a clinical assistant professor of medicine at the University of Minnesota. Her previous experience includes serving as the vice president and medical director of Consumer Health at Blue Cross and Blue Shield of Minnesota, where she was responsible for case management, disease management, and consumer engagement. She has served on the board of MN Community Measurement and the board of NorthPoint Health & Wellness Center, a federally qualified health center and part of Hennepin Health. She was named 1 of the 100 Influential Health Care Leaders by Minnesota Physician magazine in 2004, 2008, and 2012. Since 2012, she has participated in the Process Redesign Advisory Group for the National Center for Inter-professional Practice and Education coordinated through the University of Minnesota. Recently, she became a Senior Fellow, HealthPartners Institute for Education and Research. She participated in several Technical Expert Panels for Centers for Medicare & Medicaid Services (CMS) on population health measures (2015–2016), and was a member of the Population-based Payment Workgroup of the Healthcare Payment Learning and Action Network (2015–2016). She is also on the Interdisciplinary Application/Translation Committee of the Interdisciplinary Association for Population Health Sciences. She earned her bachelor's degree in pharmacy from the University of North Carolina. Dr. Magnan holds an M.D. and a Ph.D. in medicinal chemistry from the University of Minnesota, and is a board-certified internist.

Surili Sutaria Patel, M.S., is the senior program manager, Environmental Health, at the American Public Health Association. In this capacity, Ms. Patel oversees the environmental health portfolio, which includes climate change, transportation and active transportation, working with tribal governments, healthy housing, chemical safety, building partnerships, and more. She received her B.S. in political science from the University of Maryland, Baltimore County, and her master's in biomedical science policy from Georgetown University.

Jonathan Patz, M.D., M.P.H., is the director of the Global Health Institute at the University of Wisconsin–Madison. He is a professor and the John P. Holton Chair in Health and the Environment, with appointments in the Nelson Institute for Environmental Studies and the Department of Population Health Sciences. For 15 years, Dr. Patz served as a lead author for the United Nations Intergovernmental Panel on Climate Change, the organization that shared the 2007 Nobel Peace Prize with Al Gore. He also cochaired the health expert panel of the U.S. National Assessment on Climate Change, a report mandated by Congress.

Dr. Patz is committed to connecting colleagues from across campus and communities around the world to improve health for all. He is continually striving to integrate his research into teaching for students and communication to policy makers and the general public.

Dr. Patz has written more than 90 peer-reviewed scientific papers and a textbook addressing the health effects of global environmental change, and co-edited the five-volume *Encyclopedia of Environmental Health* (2011). Most recently, he co-edited *Climate Change and Public Health* (2015,

Oxford University Press) and is leading a massive open online course called Climate Change Policy and Public Health. He has been invited to brief both houses of Congress and has served on several scientific committees of the National Academy of Sciences. Dr. Patz served as founding president of the International Association for Ecology and Health.

In addition to directing the universitywide Global Health Institute, Dr. Patz has faculty appointments in the Nelson Institute, Center for Sustainability & the Global Environment (SAGE), and the Department of Population Health Sciences. He also directs the National Science Foundation (NSF)-sponsored Certificate on Humans and the Global Environment (CHANGE). Dr. Patz is double board certified, earning medical boards in both Occupational/Environmental Medicine and Family Medicine. He received his M.D. from Case Western Reserve University and his M.P.H. from Johns Hopkins University.

Celia Quinn, M.D., is the executive director of the Bureau of Healthcare System Readiness (BHSR) in New York City's Department of Health and Mental Hygiene (DOHMH) Office of Emergency Preparedness and Response. She has served in this role since 2015. The BHSR mission is to support the health care system to respond safely and effectively in all emergencies. She is board certified in pediatrics, having completed residency training and a chief residency year in the Residency Program for Social Medicine at the Montefiore Medical Center in the Bronx; this multidisciplinary training program is focused on social determinants of health, primary care, and advocacy. After completing her pediatric training, Ms. Quinn joined the Centers for Disease Control and Prevention (CDC) as an epidemic intelligence service officer and was assigned for 2 years to the Ohio Department of Health. She is currently a CDC career epidemiology field officer assigned to NYC DOHMH, and is a lieutenant commander in the U.S. Public Health Service.

Linda Rudolph, M.D., M.P.H., is the director of the Climate Change and Public Health Project Center for Climate Change and Health at the Public Health Institute's (PHI's) Center for Climate Change and Public Health. She is also the principal investigator on a PHI project to advance the integration of Health in All Policies in local jurisdictions around California.

Previously, Dr. Rudolph served as the deputy director for Chronic Disease Prevention and Public Health Promotion for the California Department of Public Health's (CDPH's) Center for Chronic Disease Prevention and Public Health and as the health officer and public health director for the City of Berkeley, California. While at CDPH, Dr. Rudolph chaired the Strategic Growth Council Health in All Policies Task Force and the California Climate Action Team Public Health Work Group.

Dr. Rudolph has also been the chief medical officer for Medi-Cal Managed Care, medical director for the California Division of Workers' Compensation, executive medical director for the Industrial Medical Council, staff physician in the CDPH Occupational Health program, and a physician for the Oil, Chemical, and Atomic Workers' International Union. Dr. Rudolph holds an M.P.H. from the University of California, Berkeley. She received her M.D. and clinical training in pediatrics and emergency medicine from the University of California, San Francisco. She is board certified in occupational medicine.

Jeffrey Thompson, M.D., is the executive advisor and chief executive officer (CEO) emeritus at Gundersen Health System. Dr. Thompson is a trained pediatric intensivist and neonatologist, and served as Gundersen's CEO from 2001 to 2016. After completing his professional training in 1984, Dr. Thompson came to Gundersen with a desire to care for patients and be a leader among his peers. He served on Gundersen's boards beginning in 1992 and was board chair from 2001 to 2014. Dr. Thompson also served as the executive vice president from 1995 to 2001 and played a key role in the organization's negotiations and governance design.

A founding member and past board chair of both the Wisconsin Collaborative for Healthcare Quality and the AboutHealth™ network, and a White House Champion of Change, Dr. Thompson has led Gundersen's nationally recognized initiatives for patient care, quality improvement, and sustainability. Dr. Thompson has certifications in pediatric critical care, neonatal and perinatal medicine, and pediatrics. He received his medical training at the University of Wisconsin Medical School, University of California, Davis, and Upstate Medical Center in Syracuse, New York. Dr. Thompson has authored and been featured in a number of articles, book chapters, and abstracts on many health care, leadership, and sustainability topics.

Fletcher Wilkinson, M.S., is the climate change program coordinator at the Institute for Tribal Environmental Professionals (ITEP). He has experience with climate and energy policy at the local, state, tribal, and federal levels. Alongside his role at ITEP, he works with tribal and rural governments in the southwest on energy issues, including renewable energy development and policy creation. He holds a B.S. in Environmental Studies and a master's in Climate Science and Solutions from Northern Arizona University.